No. 660
$7.95

FM Stereo/Quad Receiver Servicing Manual

By Joseph J. Carr

TAB BOOKS

Blue Ridge Summit, Pa. 17214

Contents

When Major Edwin Armstrong introduced his initial frequency modulation schemes in the early 30s, there was little hope of commercial broadcast exploitation of the new techniques. This was due, primarily, to the complexity of the initial system. Today, frequency modulation is widely used as a broadcast medium, as the primary modulation system for land-mobile and marine communications, and as a means of telemetry transmission.

There are actually two distinct types of angular modulation generally grouped under the term "FM." One of these is true FM in which the frequency of a carrier signal is varied by the modulating signal. The other form of angular modulation is called phase modulation. Phase-modulated signals can be demodulated by FM detectors, since this form of modulation bears many of the same characteristics as FM. The only real difference is that in a phase-modulation system it is the phase angle of the carrier signal that varies in response to the audio signal. In this system the absolute frequency of the carrier remains virtually the same.

The essential characteristics of FM and PM are pretty much the same. Frequency modulation does require a 6 dB per octave rising (with frequency) characteristic audio-frequency response (preemphasis) while PM does not. The preemphasis is needed in FM to improve the signal-to-noise ratio for high frequencies (most noise is high-frequency energy) at the receiver end of the system. This extra preemphasis is not needed in phase modulation, because the characteristic is acquired normally during the modulation process. In fact, preemphasis added to a PM system would result in an overly bright or "tinny" detected audio. To the receiver the PM signal appears like real, honest-to-goodness FM.

There are several terms and concepts that are basic to the understanding of FM in general. Although you may never work intimately with most of them, they do pop up occasionally in discussions and in printed material. Therefore, it is helpful to have a basic understanding of the meaning of each one.

One generally misunderstood term is deviation. Most people think of deviation as frequency swing. Deviation refers to the amount of percentage of modulation. It is the displacement of the carrier in kilohertz, and it is caused by the application of an audio modulating signal. Deviation is measured from the carrier center to one extreme of the signal's bandwidth. There is positive deviation as the carrier swings above its unmodulated frequency and negative deviation when the carrier swings below its no-modulation value. If the modulating signal is a sine wave, the carrier will start at the unmodulated value and increase in frequency as the sine wave signal voltage becomes more positive. When the sine wave reaches maximum in the positive direction, the maximum carrier deviation is reached. As the sine wave heads back toward the zero line, the carrier frequency decreases until it once again reaches the center or "at rest" point. When the audio sine-wave polarity reverses the carrier begins decrease in frequency. When the sine wave reaches the maximum negative value, the carrier is at the maximum negative deviation. If the modulating signal is a perfect sine wave, the negative and positive deviation percentages will be equal. If they are not equal, nonlinearity, or distortion, will be present in the audio at the receiver output.

Another frequently misinterpreted term is frequency swing, which is the total carrier displacement as measured from the maximum negative deviation to the maximum positive deviation. With sinusoidal modulating signals, the frequency swing is twice the deviation. Remember this difference. Deviation is the carrier displacement measured from the unmodulated carrier frequency to the maximum in one direction. In the United States, FM broadcasters are permitted up to 75 kHz deviation and 150 kHz carrier swing. The FCC defines 100 percent FM modulation as plus or minus 75 kHz deviation. There is no real basis for this figure. It is an arbitrary standard. In fact, TV broadcasters use plus or minus 25 kHz—one third as much—as 100 percent modulation.

If you are familiar with amplitude-modulated systems, this may seem a bit out of keeping with reality. The point where an AM transmitter becomes overmodulated is very real; it can be detected by superimposing an audio waveform on the rf carrier waveform. When the amplitude-modulated troughs reach a certain depth (baseline), we say that the rf carrier is modulated 100 percent. There is a degree of confirmation that our concept is based on reality, because several things occur rather abruptly once this point is passed. For one thing, the modulated (usually final) amplifier tube is cut off during a portion of the audio cycle. This is due to the fact that negative-going portions of the modulating signal subtract from the positive dc plate voltage used to power the rf tube. When 100 percent modulation occurs, plate voltage minus audio voltage equals zero. A corollary phenomenon, in that it is caused by the plate cutoff condition, is the generation of excessive spurious sideband components called "splatter." These components lie outside the normal AM channel bandwidth and are thus illegal. Unfortunately, the frequency-modulation limit is not so nicely defined. For the purpose of this discussion, let us say that the percentage of modulation limits were set for the purpose of standardization.

Contrary to popular belief, the amount of deviation is not dependent upon the frequency of the modulating audio in a true frequency-modulation system. However, deviation in a phase-modulated system does depend partially on the modulating frequency. The amount of deviation in an FM transmitter is determined largely by the intensity of the modulating signal. Amplitude—rather than frequency—determines this important parameter. In the phase-modulated transmitter, deviation is dependent on both the amplitude and the frequency of the modulating signal. The audio causes the phase-modulated carrier to exhibit a 6 dB per octave rising characteristic as the modulating frequency increases. This is, as stated earlier, a natural form of high-frequency preemphasis. In the frequency-modulated system, the frequency of the audio signal determines the rate at which the carrier swings between the deviation limits. It will swing completely positive and then completely negative for each cycle of the modulating audio.

The FM broadcast band was pretty much of a stepchild in the broadcasting industry for many years. Most of the early FM stations were actually affiliates of AM stations. The AM outlets were needed to pay the bills! FM simply could not compete economically with AM. Many early FM licenses were issued to stations who were motivated by the speculative nature of the venture. They felt that if the medium caught on, as it eventually did, an FM license would be a valuable property. Many FM stations transmitted the same programs as their AM affiliate station, a practice known as "simulcasting." But, in some cases, the FM owners provided exotic jazz and general classical music programing, while the AM affiliates continued to offer the normal popular fare that paid the bills.

This specialized programing became popular with serious music listeners for a number of reasons. One, of course, was that in many areas the FM band offered the only programing in classical and other so-called "fine" music. There were also some technical reasons for the increased popularity of FM for fine music. One was the wider frequency response allowed the FM broadcaster. An AM station may only transmit audio frequencies up to 5000 Hz, while an FM broadcaster is allowed to transmit audio frequencies up to 15 kHz. If this higher limit were allowed the AM broadcaster, a considerably smaller number of stations could be accommodated. The bandwidth of an AM signal is directly proportional to the frequency of the audio modulating signal.

The FM band is located in the low to medium region of the VHF portion of the radio spectrum. This wide range of frequencies and the relative "isolation" of the band has proved to be both a blessing and a curse. On one hand, the market was limited initially because of the technical complexity of the receiving equipment. On the other hand, though, the spectrum space allotted to FM provided the wide bandwidths necessary for quality sound transmission. It gave the medium exclusivity and a growing reputation for offering the best in recorded and, occasionally, live music.

Early tube-type FM tuners were a real fright. In those days, say the early to mid-50s, a typical "high-fidelity" system was, of course, monaural and boasted of a fantastic distortion figure of only one or two percent. A manufacturer who even mentioned such figures today would be drummed out of the market and given 50 lashes with a krinkled Beethoven tape if he tried to reenter! Typical FM tuners of that era were quite inferior by today's standards. This author recalls one tuner, made at home from a kit, that had the unnerving habit of changing stations as you walked across the room. If you placed your hand anywhere near the unit, the fool thing would change again. It would drift several channels just in the process of warming up! Microphonics—a ringing in the audio output caused by vibration of the elements within the tubes—were rampant. That same manufacturer, by the way, is still in business. He makes, I am happy to report, some of the best FM equipment on the market today!

Advanced engineering and the advent of solid-state circuits capable of operation in the VHF region changed the FM scene considerably. By the early 60s, the FM tuner, and the few receivers then available, were relatively reliable pieces of equipment. FM devices finally qualified for the name "high fidelity."

It was about this time (early 60s) that stereo records and tapes were beginning to make it big. Almost all new record players were equipped with an extra audio channel and a stereo pickup cartridge. In the early 60s, the FCC allowed a modification of the FM technical rules that permitted stereo broadcasting. But the Commission stipulated that stereo transmissions had to be compatible with existing standards so that no program material would be lost to the listener equipped with only a monaural receiver. The system eventually approved met this requirement nicely.

In the stereo system adopted, monaural compatibility is accomplished by linearly mixing the left- and right-channel audio signals to produce a sum labeled the L + R signal. A difference signal, needed to regenerate the discrete left- and right-channel signals from the L + R, is created by first inverting the phase of the right-channel signal 180 degrees and then mixing it with the left-channel signal. This action produces an L — R signal. The L — R signal is used to amplitude-modulate a suppressed subcarrier of 38 kHz. The result is an encoded inaudible stereo signal which is mixed with the L + R signal to decode the stereo components of the left and right channels.

Today, there are millions of stereo FM receivers in use. They range from cheapie and barely acceptable quality portables to component systems priced in the thousand dollar range and even higher. Even some FM stereo car radios are competitive with many home stereo systems where the quality, if not the power level, of the recovered program is concerned. Already, there has been much talk of using Dolby and other noise-reduction techniques to enhance the quality of an already low-noise medium . There are also several versions of four-channel, or quadraphonic stereo.

All of this activity leads to but one thing: a lot of complex and expensive equipment that will continue to be a lucrative source of work for the electronic technician. FM stereo receiver owners tend to be a nitpicking and finicky lot. They will, however, show appreciation for quality repair service on their pet music boxes by remaining loyal paying customers for years. They don't like incompetent, fly-by-night repair services, and most of them are not weasel-out-of-the-bill-type customers!

Power Amplifier
and
Driver Circuits

CHAPTER
1

The audio section is a vital link in the FM reproduction chain. After all is said and done, it is the recovered audio that interests the eventual user of the equipment. In Fig. 1-1 is a block diagram of a typical stereo receiver audio amplifier section. The input stage is dubbed a preamplifier here, but many call it a predriver to avoid confusion with the phono preamplifiers. Its function is to provide the control amplifier preceding this section of the receiver with a reasonably constant load impedance and to amplify the signal delivered by the control amplifier.

The driver is actually a small power amplifier. Its function is to provide a drive level satisfactory to the power amplifier. Of course, the power amplifier is the stage that actually develops the output signal needed to drive the loudspeaker system. In most cases, the power supply circuit is common to all three stages in the power amplifier section of the receiver.

BASIC AUDIO CIRCUITS

There are three terminals on most bipolar transistors: collector, base, and emitter. Any of these three can be used as either the input, output, or common. However, three basic configurations are normally used in audio electronics applications: common base, common collector, and common emitter.

In the common-base circuit, the input signal is applied between the grounded base and the emitter terminal. The output port is formed across the collector and base. This type of circuit is seen only occasionally in audio circuits. It is most frequently used in the rf amplifier stages in the FM tuner.

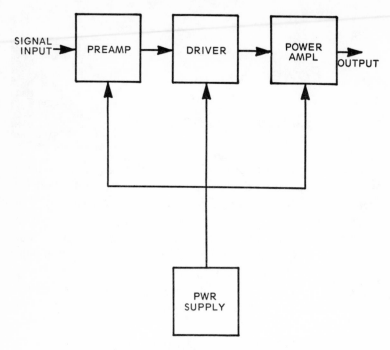

Fig. 1-1. Block diagram of a power amplifier section.

The common-collector circuit is also known as an "emitter follower." In this circuit, the input signal is fed between the collector and base. The output appears between the collector and emitter terminals. The common-collector circuit offers a high degree of stepdown impedance transformation. It is ideal for matching a high-impedance source to a low-impedance load, a condition that would normally cause distortion in the usual common-emitter voltage amplifier.

The most common circuit found in solid-state audio electronics is the common-emitter configuration. In this circuit, the input signal is applied between the base and emitter, while the output is developed across the collector and emitter terminals. The chart in Fig. 1-2 lists the various characteristics of the three amplifier types. The requirements of the overall system design determine which configuration is used in any particular case.

BIAS CIRCUITS

The circuits in Fig. 1-3 show some of the more basic methods used to bias audio amplifier transistors. One of the most simple but also least desirable bias circuits is shown in Fig. 1-3A. This is often called the fixed base current bias supply. In this circuit, a base biasing current is derived through a resistor returned from the base of the transistor to the supply voltage. Unfortunately, transistors are extremely sensitive to thermal conditions. The fixed base-current circuit makes no allowance for this phenomenon; therefore, the stage will become unstable as the voltage between the base and emitter (V_{be}) and the collector-base leakage current (I_{cbo}) vary with changes in local temperature (ambient).

In most audio circuits, several stages are connected in one of the frequently used direct-coupled cascade configurations. If this type of biasing were used in the input stage, the thermal drift of that transistor, although possibly low in itself, would be greatly amplified by the following stages. Such drift could

Amplifier Type	Voltage Gain	Power Gain	Current Gain
Common emitter	High	High	High
Common base	High	Medium	Less than unity
Common collector	Less than unity	Low	High

Impedance Characteristics		
Amplifier Type	Input	Output
Common emitter	Low	High
Common base	Very low	Very high
Common collector	High	Low

Fig. 1-2. Chart of transistor amplifier characteristics.

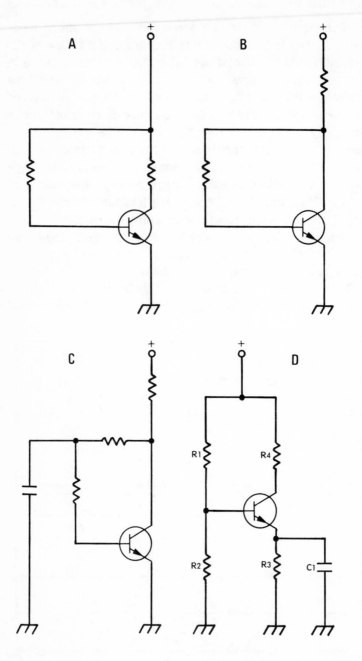

Fig. 1-3. Four circuits showing common transistor biasing methods.

easily destroy the output transistors with only a small change in the input-stage thermal conditions.

In Fig. 1-3B, we see an improved method for self-biasing. In this circuit, often called the collector feedback configuration, we have a bias resistor connected to the collector side of the collector load resistor. Since any increase in collector current (I_c) due to thermal drift will cause a lower collector voltage, we have an automatic means of reducing the base bias as the collector current tends to drift. The unfortunate part of this circuit is the problem of ac signal feedback. The collector signal is fed back out of phase to the base. This will result in a lower overall signal gain than is otherwise possible. The circuit in Fig. 1-3C is an improvement on the collector feedback idea; the result is a higher ac signal gain because of the ac feedback path decoupling to ground.

The best overall configuration is that shown in Fig. 1-3D. This circuit is a combination of fixed and self-bias. If properly designed, this circuit is the most stable. Bias is developed across a voltage divider consisting of resistors R1 and R2. In most applications, the value of R1 is substantially higher than R2. The emitter resistor, R3, provides a high degree of thermal stabilization for this circuit. R3 also increases the input impedance. This latter feature is of great use in voltage-amplifier circuits (don't believe that transistors are only current amplifiers; they also work well as voltage and power amplifiers). The collector load resistor, R4, serves the same function in this circuit as it does in the others. The absolute value of the source resistance seen by the following stage is formed by R4 in parallel with the collector impedance.

INPUT CIRCUITS

Most voltage amplifiers operate best when an infinite load impedance is coupled to an amplifier's ideal output impedance of close to zero ohms. Unfortunately, this cannot be in real life. A good compromise is achieved when a very low-impedance output is supplying signal to the very high-impedance input of the following stage. A ratio of ten to one between load resistance and source resistance is usually deemed sufficient in most applications.

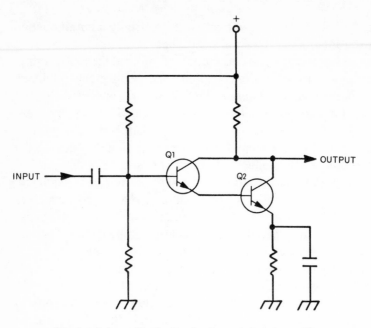

Fig. 1-4. Schematic of a Darlington pair input amplifier.

The circuits in Fig. 1-3, especially Fig. 1-3D, are often used as power amplifier input circuits. The configuration in Fig. 1-3D offers a reasonably high input impedance. The circuit in Fig. 1-4, however, is a better solution to the input impedance problem. This circuit is called the Darlington amplifier, or simply Darlington pair. Quite often, both transistors are built inside the same case with the connections already formed. In fact, integrated circuits are available with two Darlington pairs sharing a common positive dc power terminal.

There are several interesting characteristics that make the Darlington pair attractive as the input stage to an audio power amplifier: improved stability over a single stage, better gain than would otherwise be possible with an equal parts count, a much higher input impedance, and a lower output impedance.

The approach in Fig. 1-5 is often taken to improve the input-impedance situation in a power amplifier. In this circuit, a junction field-effect transistor (JFET) is used as the input stage. It is direct-coupled to a bipolar transistor. This JFET is connected in the source-follower configuration, which is favored because of its impedance-transformation properties. This type of input circuitry is even being used in certain IC operational amplifiers.

The input circuit in Fig. 1-6 has become extremely popular with audio power amplifier designers. In this configuration, we basically have a differential amplifier fed in a single-ended manner. The audio input signal is applied to the base of one transistor in the differential pair, while the base of the second transistor is held constant by a dc voltage derived from the junction of the two power transistor collector circuits. The emitters of the differential pair are fed by a constant-current source (high impedance) so that real differential action results. The collector circuits are stabilized by a pair of silicon diodes. This type of circuit exhibits low distortion and exceptional thermal stability.

FEEDBACK

Negative feedback is a tool used by all audio circuit designers. At the expense of a little overall signal gain,

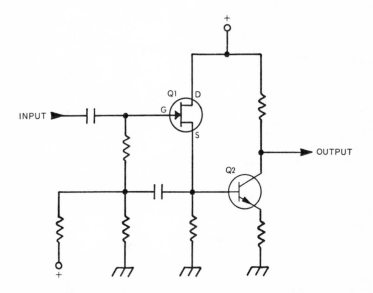

Fig. 1-5. Input amplifier circuit using a high-impedance junction field-effect transistor.

negative feedback provides better thermal and dc stability, better frequency response, and a lower overall total harmonic distortion (THD) figure. The only disadvantage, and one easily overcome, is the lower gain than is possible in the same circuit without feedback. Since a few extra stages will make up for the loss, this is not considered a handicap when compared with the benefits.

The nature of the feedback used in audio amplifiers is negative or "degenerative," since the feedback signal is out of phase with the input signal. In the typical negative-feedback system, the distortion products generated within the stage are subject to greater cancellation than the original input signal. Consequently, in the output signal there is a larger ratio between the original signal and the distortion products. This, of course, reduces the distortion products to a much lower percentage in the signal delivered to the following stages.

If there is an unevenness in the amplifier frequency response, negative feedback can help "flatten the curve." In a perfect wideband amplifier, we want a flat frequency response capability. This is to say that we want the circuit to

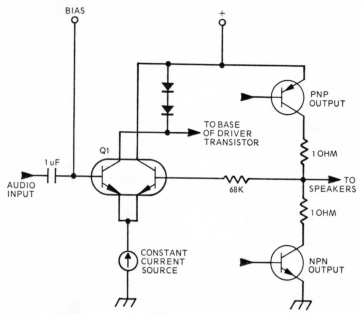

Fig. 1-6. Differential pair input amplifier circuit.

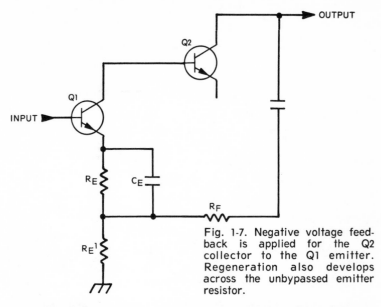

Fig. 1-7. Negative voltage feedback is applied for the Q2 collector to the Q1 emitter. Regeneration also develops across the unbypassed emitter resistor.

amplify all frequencies within the desired passband by an equal amount. Most untreated circuits, however, amplify unequally at various frequencies. In an amplifier with negative feedback, the frequencies that receive greater amplification have a greater absolute voltage level when fed back to the input. Consequently, the higher peaks in the feedback cause a higher degree of attenuation at those frequencies; in other words, there is less attenuation at "normal" signal levels than at peak levels. The result is a flatter overall response. All in all, negative feedback has to be considered a great equalizer among circuits. It provides dc stability, improved frequency response, and a lower THD figure. Negative feedback is largely responsible for the "high" part of "high fidelity."

There are numerous schemes for obtaining negative feedback in actual circuits. Two of the most common feedback circuits are shown in Figs. 1-7 and 1-8. The circuit in Fig. 1-7 is called by some the "second-collector-to-first-emitter" feedback system. It is essentially a method of voltage feedback. In Fig. 1-8, we see an alternate feedback circuit that is often called "second-emitter-to-first-base." This is a current feedback circuit.

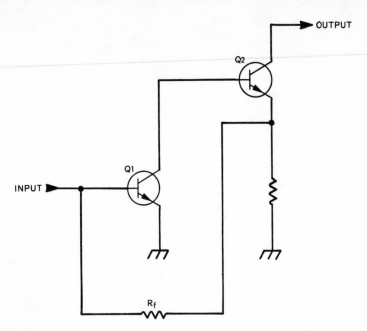

Fig. 1-8. Negative current feedback is coupled from the Q2 emitter to the Q1 base.

It is commonplace to find both systems in use in the same amplifier. In most cases, however, they appear in alternate stages. If, for instance, there are three transistors in cascade, one feedback method will be used between stages one and two, while the alternate system will be used between stages two and three. Notice that a little additional one-stage negative feedback is used in Fig. 1-7 in the form of an unbypassed emitter resistance. Leaving a certain percentage of the emitter resistance unbypassed is a common trick in audio circuitry.

PUSH-PULL OPERATION

In a push-pull circuit, two transistors are used to amplify alternate halves of the input sine wave or semi-sine wave. This type of circuit offers several advantages. One of the most important is that the push-pull circuit action causes the cancellation of all even-order harmonic distortion products. In the push-pull configuration, a pair of transistors can deliver more average power to a load than a parallel combination of the same two transistors.

One of the requirements for push-pull operation is that each transistor or transistor group must operate on alternate halves of the input cycle. This usually means that some sort of phase inversion is needed in the input. In lower-priced receivers, we might see a phase inverter circuit, such as that in Fig. 1-9, used as the driver stage immediately preceding the power amplifier. This circuit provides the collector and emitter current increases and decreases normally associated with the common-emitter circuit. As the input signal causes the collector-to-emitter current flow to increase, the voltage appearing between the collector and ground decreases. This is due to the fact that a larger share of the supply voltage is dropped across the collector resistor because of the increased current flow. At the same time, the increased current flow causes the emitter-to-ground voltage to increase. The result is a noninverted output (in-phase with the input) at the emitter and an inverted (out-of-phase) output at the collector.

The classic transformer-coupled push-pull amplifier (Fig. 1-10) is merely a solid-state adaptation of the vacuum tube configuration. Because of poorer fidelity and a higher cost to

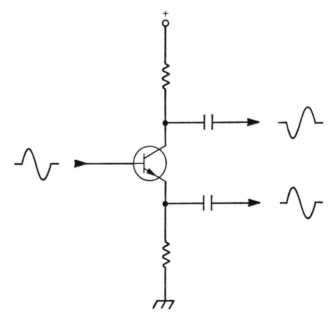

Fig. 1-9. Simple phase inverter circuit.

obtain even minimal fidelity, the traditional push-pull amplifier is of limited use in modern receivers. These circuits are found in portable radios and in some car radios, but only rarely in serious high-fidelity equipment.

Phase inversion in this circuit is accomplished in the driver transformer secondary winding. As shown in the diagram, this is a center-tapped winding. The signal sees the center tap as the common point. Therefore, the signal voltage present at the ends of the secondary are equal in amplitude but of opposite phase with respect to each other as measured against the center tap. As a result, one of the two identical transistors is driven toward conduction while the other is driven toward cutoff.

If true class B operation is used, there is no bias network attached to the center tap. Unfortunately, this is not too practical if a reasonably low value of distortion is expected. There is a phenomenon in transistor circuits called "crossover distortion." At lower values of V_{be}, no current flows in the collector circuit. The voltage level between the base and emitter must be raised to a certain minimum level before

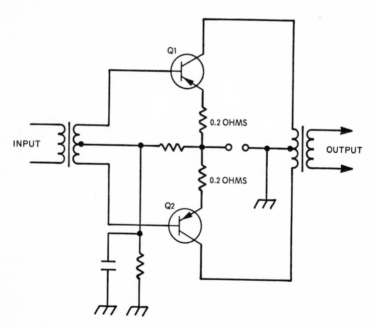

Fig. 1-10. Classic push-pull amplifier circuit using transformers.

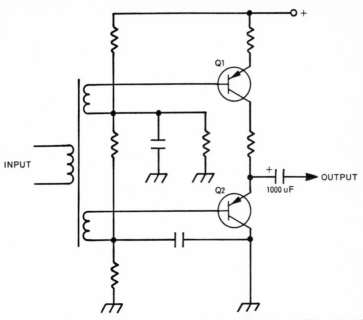

Fig. 1-11. Split secondary "totem-pole" style push-pull amplifier circuit.

conduction can begin. This causes a delay that occurs between each successive half cycle (at the points where inflection between positive and negative alternation occurs). To overcome this problem, designers give each transistor a slight amount of forward bias. Under these conditions, there is a change in collector current immediately when an input signal appears at the base terminal.

The circuit in Fig. 1-11 is a modified push-pull circuit called the "totem-pole" configuration. In this circuit, the input transformer is different and the output transformer is eliminated. Another feature of the circuit is that the two power transistors appear connected in series across the power supply. The audio input transformer has a split secondary winding which provides out-of-phase signal voltages to drive each of the two series-connected power transistors in push-pull. Coupling to the load, or speaker, is provided by a large-value electrolytic capacitor. Because of the low impedances of loudspeaker systems, this capacitor must be an extremely high value or an undesirable phase shift at low frequencies will develop. Typical values are in the 1,000 to 10,000 uF range.

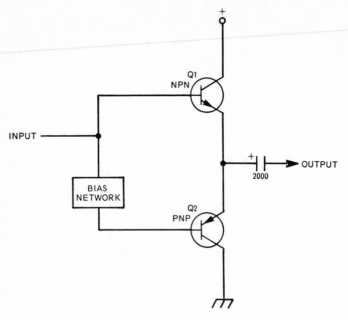

Fig. 1-12. Partial schematic of the "complementary symmetry" push-pull amplifier.

The totem-pole circuit eliminates the expensive output transformer at the cost of requiring a specialized driver transformer. But this transformer can be eliminated by the circuit in Fig. 1-12. In this configuration, we use two transistors of opposite polarity; one is an npn and the other is a pnp. The circuit is called a "complementary symmetry" power amplifier. It is critical that the electrical specifications of the two complementary transistors be identical in every important respect except for polarity.

In the complementary symmetry circuit, the opposite effects occur in each transistor as the signal voltage is applied. In the npn transistor, the collector current increases as the signal voltage drives the base terminal positive with respect to the emitter. In the pnp transistor, the opposite effect occurs as the signal voltage goes positive. As a result, the collector current decreases. It requires a negative-going signal to increase the collector current on the pnp power transistor. This means that the phase inversion needed for push-pull is accomplished automatically due to the normal differences between pnp and npn transistor types. This allows us to drive the bases in parallel; thus, the need for a driver transformer is eliminated. This helps improve the frequency response of the amplifier, because all but the most expensive transformers offer a different impedance to high, middle, and low frequencies. Both relative phase shift and frequency response are improved by this circuit at a cost generally lower than with other designs. As in the totem-pole system, the transistors are in series across the power supply, and the audio signal is supplied to the loudspeakers through a capacitor.

Unfortunately, well matched complementary pairs in the higher power classes are not always easily acquired. Low- and medium-power complementary pairs are, however, readily available. To use the complementary idea, with its inherent advantages, in a wide variety of equipment, the "quasi-complementary" circuit in Fig. 1-13 was developed. In this circuit the output transistors are connected in a totem-pole

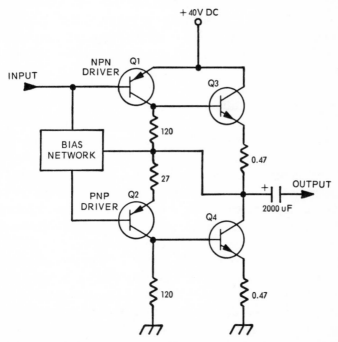

Fig. 1-13. "Quasi-complementary" circuit where the drivers are complementary and the outputs are connected in totem-pole fashion.

circuit, while the drivers are connected complementary. Thus, the advantages of complementary operation is achieved at a reasonable cost.

Some semiconductor manufacturers are offering a new type device that has found acceptance with some audio power amplifier designers. It is called the Darlington pair power amplifier device. These devices include Darlington-connected power amplifier and driver transistors, complete with driver bias resistors, housed in a type TO-3 diamond-shaped case.

A sample circuit using a pair of complementary Darlingtons is shown in Fig. 1-14. In this circuit, the complementary Darlington pairs are arranged in one of the series-connected configurations. The power transistor, driver transistor, and driver bias resistors are included inside the same package and on the same substrate to improve thermal tracking. Since all parts operate in the same thermal system, and each is made from the same piece of material, the risk of runaway and other undesirable effects due to changes in the ambient and internally generated thermal situation is markedly reduced. Transistor Q1 maintains a stable quiescent bias current. As in the previous series-connected circuits, the loudspeaker is connected through a large-value electrolytic capacitor. The bias network in Fig. 1-14 also includes any predriver transistors needed to develop the signal delivered by the control amplifier.

Many FM stereo receivers use a power amplifier that operates from two opposite polarity power supplies. They are, of course, complementary symmetry push-pull designs. An example of this circuit appears in Fig. 1-15. The one feature of this circuit that should be immediately apparent is the lack of an output coupling capacitor. As long as the power amplifier operates from two equal but opposite dc power supplies, no output coupling capacitor is needed.

Most of these circuits, and many of the capacitor-coupled circuits, use a speaker line fuse to protect expensive loudspeaker systems against a dc short in either power transistor. If this were to occur, the full current capability of the dc supply feeding that transistor would be shunted across the loudspeaker. This would undoubtedly destroy the system. The fuse cannot help the receiver in this case, but it goes a long

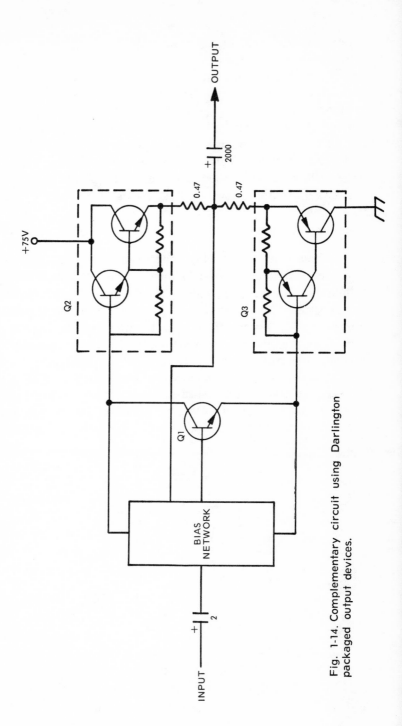

Fig. 1-14. Complementary circuit using Darlington packaged output devices.

27

way toward reducing the possibility of damage to the speakers.

The bias network in this case is shown as a block; it contains the driver and predriver transistors necessary to supply the signal to the power amplifier stage. The bias network is needed in all complementary-type output configurations. With the use of a dual power supply, it is feasible to use such circuit features as integrated circuit operational amplifiers and so forth. The use of a dual power supply also allows a greater possible voltage swing in the output signal.

PROTECTION CIRCUITS

The one irritating thing about transistor circuits, from a service technician's point of view, is the utter swiftness of their self-destruction under adverse conditions. In fact, transistors have been known to fail during near-normal operation. To overcome some of the more obvious causes of damage, manufacturers resort to certain safety measures not always deemed necessary by tube circuit designers.

Several protection techniques are shown in Fig. 1-16. There are actually three forms of fusing shown in this partial power amplifier circuit. One, of course, is the speaker fuse. Another is the dc power supply fuse. In some cases, these fuses

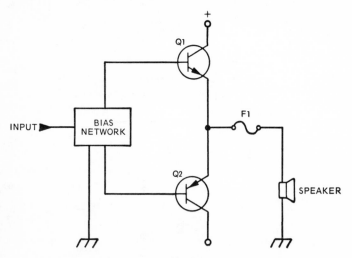

Fig. 1-15. Dual-voltage complementary output circuit.

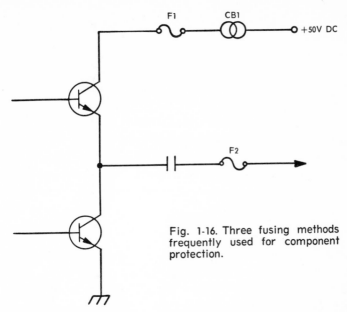

Fig. 1-16. Three fusing methods frequently used for component protection.

are fast-acting picofuse types, while in others they are "slow-blow." Why the difference? It seems to depend on how close to the normal current level the fuse vale is. It is common to find 2A slow-blow fuses in circuits similar to applications calling for 5A picofuses. It is best to replace a blown fuse with an identical type unless the equipment manufacturer recommends a larger type.

The third safety device in the circuit in Fig. 1-16, designated CB1, is a thermal circuit breaker. Circuit breakers are being used more and more in high-power amplifiers. Usually, thermal breakers are mounted on the heatsink serving one of the power transistors. In this location the breaker's sensitive metal pickup surface is at the same temperature as the transistor case. If the total ambient heat present on the heatsink exceeds a certain preselected level, the circuit breaker will open and thereby remove the dc power from the amplifier. In some cases, a latch circuit is provided to turn on a front panel warning light to let the listener know why his music suddenly disappeared.

Overcurrent can kill a solid-state power amplifier in a hurry. A circuit designed to prevent damage caused by a sudden increase in current is shown in Fig. 1-17. In this circuit,

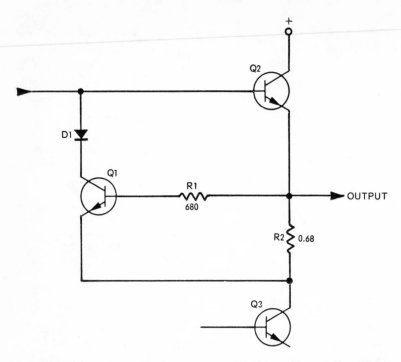

Fig. 1-17. Some circuits use a protection transistor in the output stage.

Q2 and Q3 are the normal stacked power amplifier transistors. For the sake of simplicity the drivers and predrivers have been omitted. Transistor Q1 is the protection transistor. (In actual practice there is another transistor to protect the other power transistor.) If the Q2 collector-emitter current rises, the dc voltage drop across the 0.68-ohm resistor will rise also. If the current becomes excessive, the dc voltage drop will increase to a point high enough to forward bias protection transistor Q1.

In some sets, this transistor receives a slight amount of forward bias from other sources so the drop across R2 need not rise very high. In other cases, however, the only forward bias needed is that obtained across resistor R2. When the dc level across R2 rises to the threshold voltage, Q1 is forward biased and saturated. This drastically lowers the resistance between the Q1 collector and emitter. In effect, this resistance shorts out the Q2 base-emitter junction by removing the forward bias. This, of course, turns off the transistor.

OTHER CIRCUIT FEATURES

One method used to obtain the forward-bias level necessary to overcome crossover distortion is shown in Fig. 1-18. A diode (or diode stack in some cases) is used to supply the bias. You may wonder why the diode doesn't clip one half of the input cycle. The answer lies in the nature of diode operation. Solid-state diodes will pass ac signal voltages that are small compared to the level of dc forward-bias current flowing. The ac will pass superimposed on the dc level.

You cannot replace such diodes with a single unit until you are sure of the precise nature of the old diode. In many cases, the diode may actually be a stack of several diodes in a package normally used to house a single diode. In any event, the dc quiescent voltage drop must be some multiple of the normal 0.6V dc contact potential of a single diode. If the drop is 2.4V dc, for example, assume that the "diode" is actually a stack of four silicon diodes in one epoxy package.

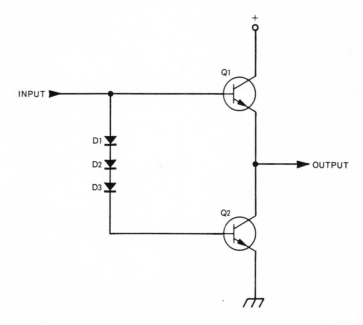

Fig. 1-18. Diode forward biasing eliminates crossover distortion in a push-pull stage.

The circuit in Fig. 1-19, frequently encountered in audio output stages, may seem a bit out of place at first glance. Several of thse parts, as well as several parts in the driver and predriver stages, seem better suited to rf applications. The rf choke (RFC1) and the shunt 680 pF capacitor suppress any possible oscillation in the amplifier circuit.

This tendency is the result of several conspiring factors. One is the inherent high gain needed from the power amplifier. Another is the possibility of positive feedback at rf frequencies due to stray wiring capacitance. This problem is compounded by the fact that many of the transistors used in audio amplifiers are responsive to high-frequency rf currents.

It is a well known phenomenon that the practice of cascading stages reduces frequency response. That is to say that if you cascade a number of identical stages, the overall frequency response will be less than that of any single stage. The more stages in a chain, the greater the reduction in overall bandwidth. With three identical stages, the amplifier will offer approximately one half the frequency response of

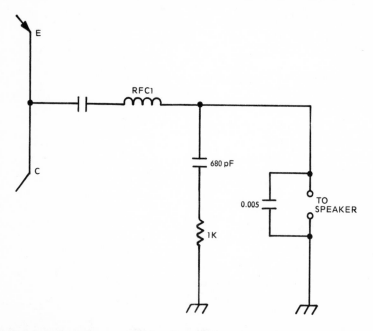

Fig. 1-19. The rf choke and small-value bypass capacitor prevent high-frequency oscillation in the output stage.

any one stage. Since modern amplifiers are said to be capable reproducing frequencies up to 10 kHz, you can see why they must use transistors with high-frequency specifications in order to insure the claimed bandwidth.

Ultrasonic oscillations cause distortion and eat up power that could otherwise be used to drive the loudspeaker system. The presence of "ringing" oscillation is indicated by blurs on the trace of a square wave viewed on an oscilloscope. The listener in these cases often complains of a "lisp," especially when the announcer or vocalist enunciates an "esss" sound.

Control Amplifiers and Preamps

The control amplifier is responsible for shaping the overall characteristics of the audio portion of the receiver and for much of the so-called ease of operation offered to the user. All of the normal audio controls are included in this block: volume, loudness, treble, and bass. In most cases, the control amplifier is mounted along the front edge of the receiver chassis so that the control shafts extend through the front panel.

The preamplifier stages are generally used only on the phono function. Most of the higher quality phonograph pickups used today are of the magnetic variety. Although they are capable of extremely good performance, frequency response-wise, they provide a very low-level output signal. An output voltage of ten millivolts (10 mV or 0.010V) would qualify a magnetic cartridge as a regular dynamo. Most cartridges offer outputs in the 5 mV range. With such a tiny signal, a preamplifier is needed to build this signal up to the level of the control amplifier input requirements. The preamplifier circuits are designed to provide a flat-response output from the phono or tape head input.

The block diagram in Fig. 2-1 shows the stages included in the control section of a good quality stereo receiver. For the sake of simplicity only one of the audio channels is shown. Since we are assuming that the equipment is of high quality, the tone control is the type that varies the amount of negative feedback used in the amplifier. The volume control is usually found between two control amplifier stages. This is not, however, an absolute rule. The loudness control is usually switched and most probably operated from a tap on the regular volume control. This switch modifies the frequency

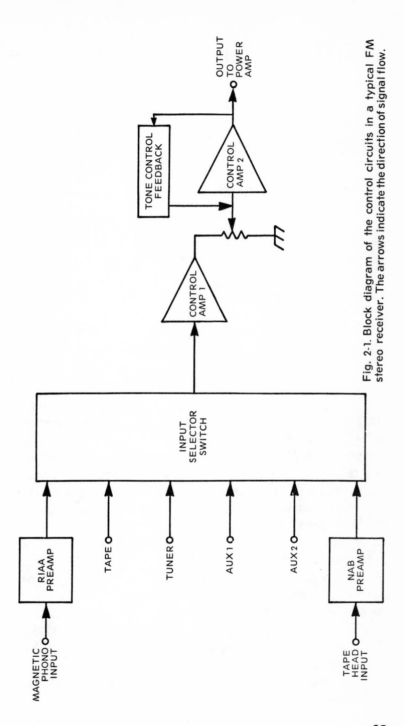

Fig. 2-1. Block diagram of the control circuits in a typical FM stereo receiver. The arrows indicate the direction of signal flow.

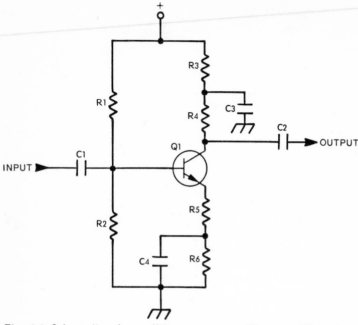

Fig. 2-2. Schematic of a voltage common-emitter amplifier stage.

response of the system at low volume levels so that it more nearly corresponds to the hearing ability of the human ear.

The input to the control amplifier is selected by a rather awesome switching network. In most receivers, this function is handled by either a multiple-wafer rotary switch or a complex bank of pushbutton "lock-out-all-but-one" switches. The tape and tuner inputs generally go through the switch system with only the necessary impedance-matching network. If the input impedance to the control amplifier is high enough, this is simply a simple resistor or RC network. In a very few cases, there is a special "tape head" input, which goes to a preamplifier circuit that offers the standard NAB tape playback equalization. In most designs, however, the tape function is handled like the other two inputs without any special preamplifiers. Any auxiliary inputs are of the "straight through" or wideband variety. In the phono position, however, some special circuitry is necessary. In that position, the response curve must provide RIAA equalization, and the extra voltage gain needed requires the use of at least one additional stage.

TYPICAL CIRCUITS

The circuit in Fig. 2-2 is what might be called a typical single-stage audio voltage amplifier that can be used in either control or preamplifier sections. This common-emitter circuit is pretty straightforward with no fancy tricks used to compensate for one defect or another. A slight amount of negative feedback is obtained by omitting the bypass across part or the emitter stabilizing resistance. The degree of emitter bypassing determines the response characteristic of the stage. In general, the circuit bandpass is relatively flat if the reactance of C4 is less than one tenth the resistance of R6 and if the ratio R6 over R5 is large.

Another method is used to improve the frequency response in this circuit (Fig. 2-2). The load in the collector is split and a bypass capacitor is connected to ground across part of the collector resistance. The remaining amount of tone shaping depends on the values of the signal-coupling capacitors. If the reactance of these parts is low compared with the normal impedances in their respective circuits, the amplifier will be essentially a wideband type. If, on the other hand, the capacitive reactance is an appreciable part of the impedance, the circuit will tend to restrict the low-frequency bandpass of the amplifier. This circuit device might possibly be found in certain preamplifiers but not in wideband control amplifiers. Although the parts count is increased slightly, there is something to be said for the "bootstrap" audio amplifier in Fig. 2-3. In this configuration, the collector characteristics of the input stage are designed to match the base characteristics of the following stage. Although the I_{cbo} leakage can cause an appreciable operating-point drift problem in this circuit, proper design will reduce the severity to almost nothing.

The absence of series coupling capacitors between the two stages enhances the low-frequency response and reduces the degree of overall phase shifts. Added to these advantages is the extra gain allowed by the "bootstrap" configuration. This circuit is capable of developing more gain than a single transistor can produce without the necessity of having each device work to its maximum ability.

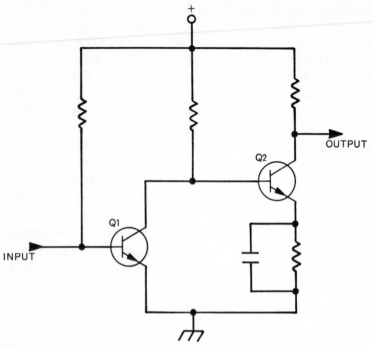

Fig. 2-3. Schematic of two audio amplifier stages in a direct-coupled cascade arrangement.

The operational amplifier, or simple "op-amp," is a unique building block for audio equipment. Originally, the op-amp was used to perform mathematical operations in analog computers. In the world of modern industrial and defense electronics, op-amps are used in computer applications. To describe an operational amplifier, we can say that it is a gain block with an infinite input impedance, an infinite open-loop (no feedback) gain, and a zero output impedance. In reality, these "paper" specifications are never realized. Available IC op-amps have an input impedance of a megohm or more, an open-loop gain (A_{fol}) of over 100,000 (approaching 1,000,000 in some), and an output impedance of less than 100 ohms.

The key to the operation of an op-amp is the feedback loop. This is shown in Fig. 2-4. An impedance is connected between the inverting or minus input and ground. The ratio of the feedback impedance to the ground impedance determines the closed-loop gain of the stage. To realize a gain of ten (20 dB), the value of the feedback resistor must be ten times greater than the ground resistor. If a gain of 1000 (60 dB) is needed, the feedback resistor value must be 1000 times greater than the ground resistor.

Since this can lead to either a low input impedance or instability (depending upon the op-amp), designers often use other tricks to raise both the gain and the input impedance. One such trick is to connect the input resistor to the negative input in series with the signal path. A typical value for this resistor might be on the order of 10K. A 1M feedback resistor is returned to a voltage divider across the output. The value of the ground resistor from the positive input must be equal to the feedback resistor rather than the other input resistor; otherwise, capacitor coupling is needed. The only disadvantage with an op-amp is that it must be operated from either a dual voltage supply or from a single-ended supply with dropping resistors and zener diodes arranged to simulate a dual voltage supply.

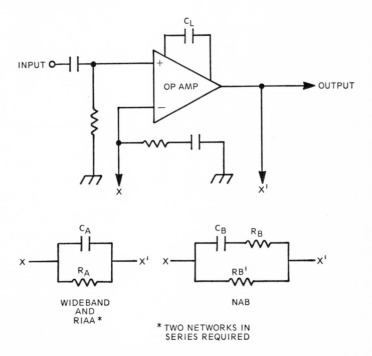

Fig. 2-4. Operational-amplifier audio circuit with feedback equalization networks.

The most common op-amp used in audio equipment is the dual type shown in Fig. 2-5, which is the Motorola MC1303P. A number of other op-amps by Motorola and others are also used. Some may be interchangeable with the type shown, while others are entirely unique.

The op-amp, especially in its dual-package configuration, has found popularity among stereo designers because it allows them to get good results with little extra effort. The op-amp permits the use of the building block concept. It is possible to extend this philosophy to a point where a manufacturer needs only one basic printed-circuit board foil pattern. This means that the same board can be used for a wide variety of audio amplifier applications with only minor variations in component values or configuration. Obviously, such a printed-circuit board has certain redundant patterns in most applications, but it can serve as the basic amplifier board in all stages up to the predriver.

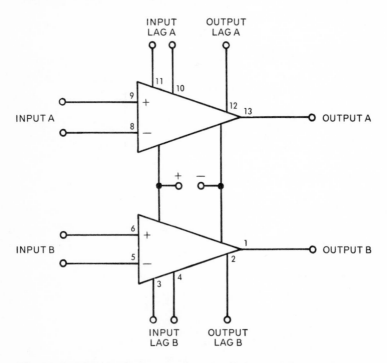

Fig. 2-5. Input and output connection of a Motorola MC1303P IC dual op-amp used in stereo circuits.

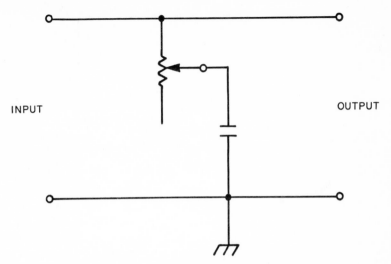

Fig. 2-6. Simple treble rolloff type tone control circuit.

TONE CONTROLS

In the absolutely lowest-priced stereo receivers, we find only the simplest type of tone control. This is the so-called "treble rolloff" type used in cheap table model radios. In this circuit there is a single control for both bass and treble. A typical treble rolloff tone control circuit is shown in Fig. 2-6. The control circuit is composed of a single potentiometer and a fixed capacitor. If the capacitor were placed directly across the audio line, there would be severe attenuation of high frequencies. By varying the potentiometer setting, we can determine the effect of the capacitor across the line. There are two disadvantages with this type of circuit. It severely cuts the volume of treble frequencies (hence the name "treble rolloff") and it produces a poorly controlled artificial tone.

Many years ago when the vacuum tube was still used in quality hi-fi equipment, the most popular type of tone control was called either the "boost" system or the "series-shunt" system, depending upon the manufacturer consulted. An example of this style tone control circuit is shown in Fig. 2-7. Although this tone control circuit was used extensively in tube designs, it is seen only occasionally on low- to medium-priced receivers today.

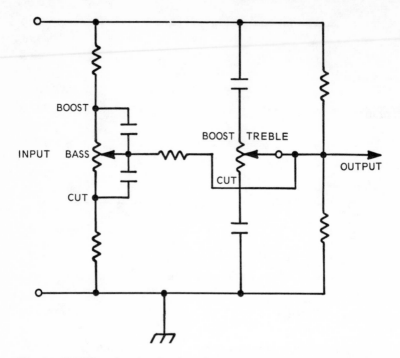

Fig. 2-7. Traditional series-shunt or "booster" type tone control circuits.

The most common type of tone control circuit used in modern equipment is the Baxandall or "negative-feedback" circuit shown in Fig. 2-8. In this arrangment, two or more RC passband filters are placed in the negative-feedback loop of the amplifier stage. Negative feedback lowers the gain of an amplifier. If the circuit is selective about the frequencies passed by the feedback loop, the amplifier is selective about the frequencies that it passes. If, for example, you wanted an amplifier to act as a high-pass filter with a certain cutoff frequency, you could insert a low-pass filter with the same cutoff frequency in the negative-feedback loop. In this situation, only those frequencies below the cutoff point would reach the input of the amplifier from the output via the feedback path. As a result, these frequencies would be attenuated in the amplifier by the suppressing action of the feedback. Higher frequencies, those that you want the amplifier to pass, will be unaffected. The circuit simply appears as another wideband gain block to those frequencies above the cutoff point. If the passband filters placed in the feedback loop of the control stage are designed with a sufficient overlap of bass and treble action, the control over tonal characteristics will be continuous throughout the entire audio spectrum.

An offshoot of the Baxandall tone control circuit is the tone-equalizer concept. In this arrangement you might find a receiver with up to ten separate tone controls, where each covers a band of frequencies within the audio spectrum. In these circuits, the control elements are a series of resonant circuits (series tuned) that offer a low impedance at the resonant frequency and those frequencies close to resonance. At one extreme of the control potentiometer, the trap is shunted across the audio input line. This introduces attenuation at the trap's resonant frequency. The degree or amount of attenuation is determined by the position of the control wiper. At the opposite extreme of the control, the series resonant trap acts as a shunt to ground in a negative feedback network and prevents those frequencies close to

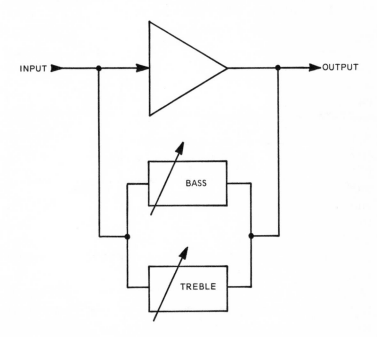

Fig. 2-8. Block diagram of the Baxandall or negative-feedback tone control circuit.

resonance from traversing the feedback loop. The result is increased amplification of those frequencies in relation to the remainder of the audio spectrum.

SPECIALIZED EQUALIZATION

The output of phonograph cartridges and tape player heads is not flat. Because of this, there is a need for a certain amount of tonal modification to make the reproduced sound more believable. Typical equalization curves are shown in Figs. 2-9 and 2-10. In Fig. 2-9A we see the RIAA standard curve for phonograph systems. This curve is easily identified by the two flat zones and two zones of descending frequency response. Curve B represents the response of a typical wideband amplifier such as might be used in the control amplifier section of a stereo receiver. This curve completely eliminates any coloration of the program material except at the ends of the passband. This curve says that the amplifier is essentially flat in its total frequency response.

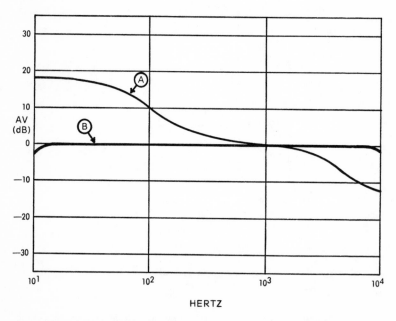

Fig. 2-9. RIAA and wideband frequency response curves.

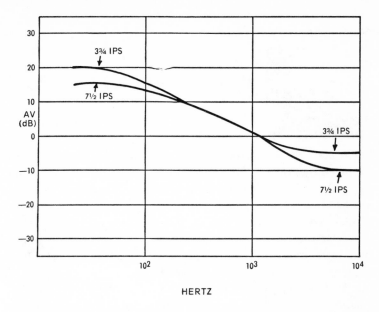

Fig. 2-10. NAB frequency response curves for 3¾ and 7½ ips tape speeds.

The NAB tape equalization curves in Fig. 2-10 illustrate the output response of magnetic tape systems. Since tape speed affects the playback frequency response, there is a slightly different curve for each of the two speeds commonly found on home tape machines: 3¾ ips and 7½ ips. In most receivers, there probably isn't one single preamplifier with these characteristics unless there is a special tape head input. The usual tape input on FM stereo receivers is intended for high-level, wideband signals that can be fed directly to the control amplifier via the selector switch. The tape player manufacturer supplies the equalization in the line output circuit of his machine. His equipment is expected to deliver an essentially flat audio response to the tape input of the receiver.

CHAPTER 3

RF and Detector Circuits

Little success was predicted for FM when it was first introduced by Major Edwin Armstrong in 1933. It was believed to be too complex for ordinary home use. Time and the subsequent engineering work in this field have proved the early critics wrong. Today, FM is the lifeblood of high-fidelity broadcasting. An FM station is allowed a greater audio frequency bandwidth (a higher "fi"); its signals are less susceptible to "skip" interference at night due to the nature of the VHF signal; and FM reception is relatively free of atmospheric and certain types of man-made interference. This set of happy circumstances gives FM a much better all-around reputation than AM. The AM band still has its place and use, of course, but for high-quality broadcasting FM is king.

In AM it is the amplitude of the radio signal that is varied to convey the program material. But there are certain limitations that detract from AM as a high-fidelity medium. For one thing, static-type noise interference tends to apply additional amplitude modulation to the signal being received. If you take steps to get rid of the objectionable noise, you lose a portion of the program material. In FM, it is either the absolute frequency or the phase of a fixed frequency that is varied according to the audio program material. Since noise amplitude-modulates the signal, we can clip the peaks of the carrier in a circuit called the limiter and remove the interference without disturbing our program.

We can't say that FM is "noise free," only that it is less susceptible to noise. If the strength of the signal is high enough to be above a receiver's sensitivity limit, the listener can expect noise-free reception. However, if the signal level is lower than the sensitivity limit, some noise may be heard. But, lower than the sensitivity limit, some noise may be heard. But, since most FM detecotrs respond only poorly to amplitude modulations, this noise will probably be less severe than it would be on an AM receiver. Also, the effect of noise is less noticeable because the FM carrier is located in the VHF region where the harmonic amplitudes of noise signals are noticeably weaker.

DETECTORS

There is a wide variety of available FM detector circuits, although it must be said that this has not always been the case. Many of today's circuits are redesigns of older types that were considered too difficult to adjust.

There are several criteria against which we can measure the performance of an FM detector. One of these is that it must be capable of extracting the audio program information contained in either frequency or phase variations from a center frequency. Secondly, it must perform this function over a relatively wide dynamic (amplitude) and frequency response range. It must also be stable so that it does not require frequent adjustment by either user or service technician.

Discriminators

Some years ago, the Foster-Seeley discriminator was designed to eliminate some of the difficulties inherent in the older Travis discriminator. The Travis discriminator had three tuned circuits which had to be balanced in critical phase relationships. The original Foster-Seeley circuit was designed to operate with vacuum tube diode rectifiers. However, current F-S circuits are using solid-state diodes. The only real disadvantage of the Foster-Seeley circuit is that it is not the best at suppressing amplitude modulations on the carrier. Because of this, it is always preceded by a limiter stage or stages designed to clip the amplitude peaks of the signal. This action helps insure a clean audio signal.

The circuit in Fig. 3-1 is a simplified but almost typical Foster-Seeley discriminator. In this circuit, we obtain demodulation from the phase relationships that exist within the tuned transformer. The primary of the detector transformer is series connected (rf reference to ground) with the secondary winding. The signal voltages in this winding are 90

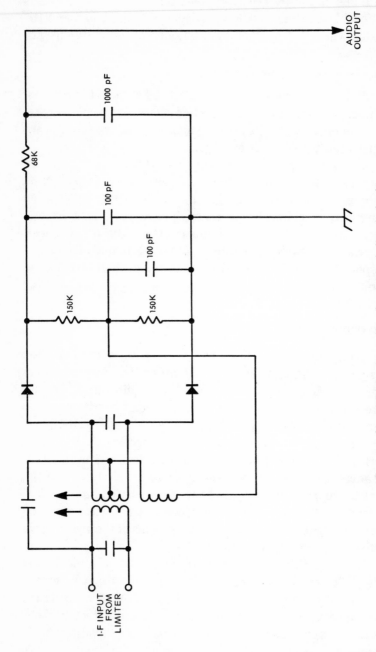

Fig. 3-1. Simplified schematic of a typical discriminator.

I-F INPUT FROM LIMITER

AUDIO OUTPUT

1000 pF

100 pF

100 pF

150K

150K

68K

degrees out of phase with the voltages existing across the secondary as a whole and 45 degrees out of phase with the voltages in each half of the secondary.

When a steady signal at the carrier center frequency is present across the input, equal but opposite signal voltages appear at each diode. Consequently, the dc resultant across the diode load resistors is zero. When the input signal is modulated, however, the situation changes. The signal voltage is greater or less first at one diode then the other. This results in a changing dc level across the load resistors. The change is proportional to the audio modulation on the rf carrier wave. Appropriate capacitor output coupling applies the recovered audio to the following audio amplifiers.

The specifications of the parts that make up the discriminator are rather critical. First, and probably foremost, the transformer must be of the special type designed for this type of service. The diodes must be closely matched. This is not a tight and rigid requirement for most lesser equipment, but for high-fidelity equipment it is mandatory. Most manufacturers can supply prematched diodes at a slight premium over the price of two single diodes. The diode load resistors should also be closely matched. This means the use of at least five percent and preferably one percent tolerance components.

Ratio Detector Circuits

In Fig. 3-2 is the basic ratio detector circuit. Although this is a newer circuit than the Foster-Seeley, it is still considered an "old hand" in the FM demodulation business. The ratio detector is far less sensitive to amplitude variations, most of which are smoothed out by C3, so it need not be preceded by a limiter. This circuit closely resembles the Foster-Seeley discriminator. On closer inspection, however, you will notice that the polarity of one diode is reversed, while in the discriminator they face the same direction. In the ratio detector circuit, the diodes are connected so that the output voltages add rather than subtract as in the discriminator.

When the unmodulated rf signal is impressed across the transformer primary, the relative contribution of each diode

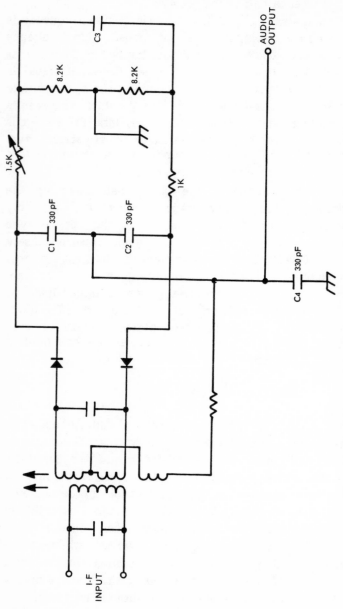

Fig. 3-2. Simplified schematic of a typical ratio detector circuit.

to the output dc level is a ratio of two to one. This ratio changes though, when the input signal is frequency modulated. In the latter case, the ratio goes up or down, depending on the deviation and deviation rate. The important thing to remember is that a modulated input signal causes the ratio of each diode's dc contribution to the output to vary from the steady-state value of two to one.

In some ratio detector designs there is a potentiometer in one leg of the circuit to compensate for the differences in dc diode characteristics. Even though this circuitry gives us some latitude in the selection of replacement diodes, it should still be standard practice to replace both and use a matched pair.

Capacitor C3 is known as the AM suppression capacitor, although its function is actually twofold. First, the capacitor does help eliminate amplitude modulation; therefore, it is useful in suppressing noise. Secondly, it is used to replace the battery needed in early ratio detector designs. This capacitor continuously charges with rectified FM i-f signal, which maintains a dc level that is proportional to the incoming signal strength in many cases. Also, the voltage across the capacitor is often used as the indicator point for FM alignment.

Quadrature Detection

The introduction of integrated circuits inevitably led to the utilization of certain FM detection techniques that would have been a little on the complex side if it were necessary to use discrete components. One of these techniques is quadrature detection. A pair of signals of the same frequency are said to be "in quadrature" if there is a precise 90 degree phase difference between them with respect to time.

The block diagram of the internal circuitry of a typical quadrature detector appears in Fig. 3-3. The input signal obtained from the i-f amplifier chain is first subjected to further amplification and hard limiting in a set of wideband amplifiers designated here as A1, A2, and A3. Then, the output is split into two components. One is fed directly to the internal synchronous detector circuit (itself a complex collection of

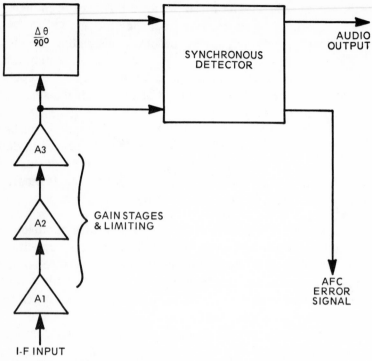

Fig. 3-3. Block diagram of an IC quadrature detector circuit.

transistors that would be too expensive without IC technology). The other component of the limited i-f signal is fed to the IC through a tank circuit tuned to cause a 90 degree phase shift at 10.7 MHz. The phase-shifted signal is applied to the other input port on the detector. These two signals are recombined vectorially in the gated synchronous detector, and the result is a variable width square-wave output pulse, which is integrated into an audio signal. This stage also produces the afc correction voltage that is needed to keep the FM local oscillator on frequency.

Some technicians prefer to align this stage with the aid of a total harmonic distortion (THD) analyzer. It is a simple matter to adjust the phasing coil until the lowest THD figure is reached with a 400 or 1000 Hz audio sine-wave signal input. This circuit, by the way, seems a little less sensitive to random static than are most diode types because of the high degree of hard limiting.

Phase-Locked Loop Detection

You will see more and more phase-locked loop (PLL) circuits in FM stereo receivers as time passes. All major semiconductor manufacturers already offer FM PLL chips. A block diagram of the basic PLL detector is shown in Fig. 3-4. The heart of the PLL detector is a voltage-controlled oscillator (VCO). The VCO output frequency changes with a variation of a certain dc input control voltage. The rf output of the VCO is fed to a phase detector. Also feeding this same phase detector is the modulated FM i-f signal. This circuit acts as a comparator to produce a dc output proportional to the phase difference between the two signals. A filter removes any residual rf component from this signal before passing it on to a dc amplifier. If the input signal is steady and unmodulated, the dc will be either steady or zero. When the FM signal frequency deviates under modulation, however, the phase comparison is upset and a dc correction voltage is generated. This voltage is

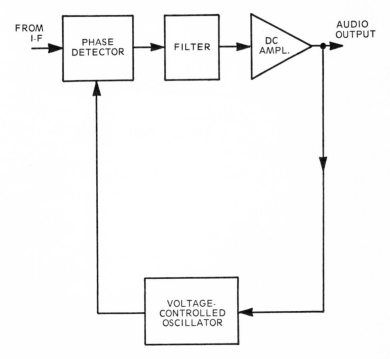

Fig. 3-4. Block diagram of a phase-locked detector.

used in an attempt to bring the VCO frequency into line. It is this changing dc level that is used as the audio output signal.

The circuit in Fig. 3-5 shows the connections to a typical PLL chip. The RC network, consisting of potentiometer R1 and C_A, determines the frequency of the VCO. The characteristics of the filter, including the range of frequencies over which the PLL will lock, is set by capacitor C_B. The impedance of the input circuitry, by the way, is low enough to be compatible with several ceramic i-f bandpass filters already on the market. Since these filters are popular with designers, you can expect to find them in many PLL detector circuits. One interesting feature, regarding the use of an IC PLL, is the possibility of also building all or at least most of the FM i-f voltage and power gain into the same chip that contains the PLL. This makes feasible a small but very high-quality FM receiver.

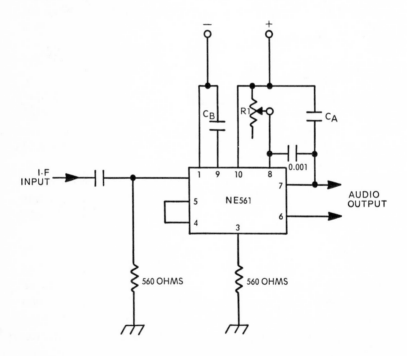

Fig. 3-5. Schematic of a phase-locked loop detector using the Signetics NE561 IC.

I-F AMPLIFIERS

The purpose of the i-f amplifier is to provide the bulk of the gain needed in the system and to limit and shape the system passband to allow only the desired input signal to pass while providing a high degree of rejection to other signals. In the undersired category are the signals of other radio stations and noise impulses falling within the FM band.

Until recently, most FM i-f amplifiers have used bipolar transistors as active elements. More modern types, however, use one or more types of integrated circuits designed especially for service as FM i-f amplifiers. A typical bipolar FM i-f stage is illustrated in Fig. 3-6. Receivers featuring this type of stage usually have three or four such amplifiers in cascade to provide the needed gain. Of these the fourth (and perhaps the third) may be a limiter.

The common-emitter configuration is normally used for bipolar FM i-f amplifier circuits. The input and output sides of the circuit are coupled to other stages through tuned transformers. In some cases, as in the circuit in Fig. 3-6, the transformers may be of the quadratuned variety. This type of transformer offers a greater degree of selectivity at the cost of a slight amount of gain. The quadratuned transformer is most likely associated with the first stage in the i-f amplifier chain or at the output of the mixer. In the case in Fig. 3-6, the input i-f transformer is capacitively coupled to the base of the transistor. In other circuits of similar configuration, a low-impedance tap on the secondary of T2 might be used for this purpose. Conventional bypassing and biasing techniques are used in this circuit.

A very common type of integrated circuit i-f amplifier is shown in Fig. 3-7. This circuit uses the well known uA703 IC as the active element. This IC is produced in one version or another by virtually every semiconductor manufacturer who offers linear ICs. Both imported and domestic versions of the IC are available. The uA703 is used in equipment by an extremely large number of manufacturers. The internal circuitry is basically a diode-stabilized and biased differential amplifier.

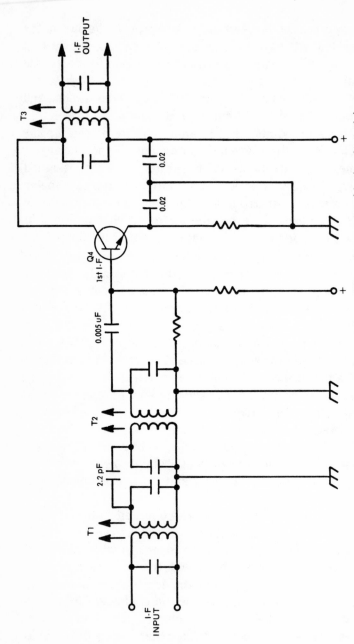

Fig. 3-6. Schematic of a typical i-f amplifier stage using bipolar tran-sistors.

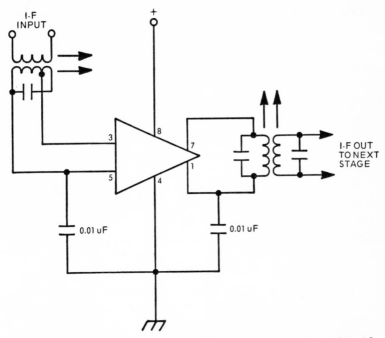

Fig. 3-7. Schematic of an i-f amplifier stage using the popular uA703 IC.

The IC has only the six leads for the following connections: positive power, negative power and signal common, differential input leads, and differential output leads. In earlier production models, the uA703 was manufactured in a plastic-epoxy case. This six-lead case is about the size of a TO-5 transistor package. More modern types, designated the uA703A, are enclosed in a six-lead version of the standard "similar to TO-5" metal can. These newer types are reputed to be more reliable than the older epoxy types. In any event, they are electrically and mechanically interchangeable with the older style in 10.7 MHz FM i-f service. In many stereo receivers, two uA703 stages are preceded by a single bipolar stage. In a few cases, there may be up to four IC stages.

TUNER

To achieve true superheterodyne reception, it is necessary to mix together—in a nonlinear stage—the incoming rf signal and a signal generated by a local oscillator. The frequency of the local oscillator is equal to the rf frequency plus or minus (usually plus) the i-f. If these signals were mixed together in a linear element such as a resistor or reactance, the result would be two signals superimposed on one another. They would, however, remain only two separate and distinct signals.

Mixer

To make a superheterodyne work, it is necessary that the rf and oscillators signals beat against each other to produce at least one additional frequency, the i-f. The purpose of the mixer circuit in a superheterodyne receiver is to provide the circumstances that will allow such mixing.

There are a number of ways in which this can be accomplished. One is to use a simple solid-state silicon diode. Many years ago, low-cost VHF receivers of low sensitivity, as well as more recent UHF TV tuners, use just that very trick to provide nonlinear mixing. Unfortunately, the diode not only provides zero gain but it actually produces a large signal loss. In the days when VHF amplifiers were either troublesome or even nonexistent, this was the only way to design a VHF receiver. Today, however, we have a wide variety of solid-state devices capable of operating as either amplifier, oscillator, or mixer at VHF.

One circuit, using a junction field-effect transistor (JFET) as the active element, is shown in Fig. 3-8. In this circuit, the incoming rf signal is coupled to the JFET gate terminal. Under normal circumstances, we could tune the drain terminal to the input frequency and this stage would function as another stage of rf amplification. In this application, however, the drain is tuned to the i-f frequency, which is usually 10.7 MHz. Under these conditions, there will be no output from the mixer unless a second signal (in the 100 MHz range) is present to convert the rf signal to 10.7 MHz. This second signal, supplied by a local oscillator, is fed to the source terminal of the JFET. The local oscillator signal voltage alternately aids then bucks the normal bias voltage on the JFET. Thus, the

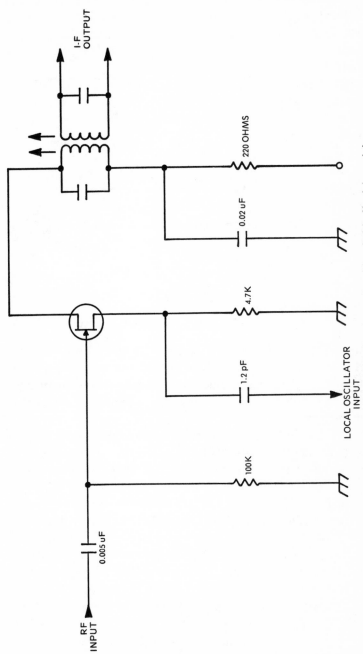

Fig. 3-8. Schematic of a mixer circuit using a junction field-effect transistor.

I-F OUTPUT

220 OHMS

0.02 uF

4.7 K

1.2 pF

LOCAL OSCILLATOR INPUT

100K

0.005 uF

RF INPUT

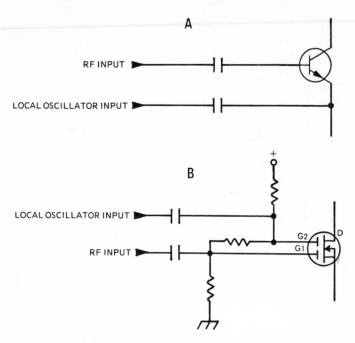

Fig. 3-9. Bipolar and dual-gate IGFET mixer circuits.

oscillator signal introduces a switching action. This is just the sort of nonlinearity that will result in the production of a heterodyne difference frequency. This difference frequency, the i-f, is coupled to the following stages via the tuned output transformer.

There are other signals present at the drain besides the i-f. These include the local oscillator signal, the rf signal, and a forth signal that is equal to fosc + Frf. These are shunted to ground via the 0.02 uF decoupling capacitor. The i-f transformer is tuned to pass only the signal that is fosc—frf.

Other solid-state devices are used as the mixer in a wide variety of receivers. Two more common circuits, shown in Fig. 3-9, use either regular bipolar transistors or the newer insulated-gate field-effect transistors (IGFET, also called MOSFET by many). The action of the bipolar circuit in Fig. 3-9A is similar in most essential details to the action of the JFET circuit just discussed. The IGFET circuit in Fig. 3-9B, however, is a bit different. The IGFET has two input gates.

One is fed by the rf signal, while the other is fed by the local oscillator signal. The local oscillator alternately "pinches off" then allows the rf signal to flow. The result is a nonlinear switching action capable of generating the i-f. These IGFETs can be a bit sensitive to handle. Older type numbers had a very sensitive gate structure that was susceptible to damage by static electricity charges on your tools and hands. Most of the more modern types, however, use internal zener diodes to shunt dangerous dc or static levels harmlessly around the delicate gate insulation.

RF Amplifiers

Field-effect transistors are also used as rf amplifiers in FM receivers. These devices have many of the characteristics of pentode vacuum tube amplifiers, including high gain, high input impedance, low internal noise, and a superior ability to handle intermodulation and other forms of signal overload. Figure 3-10 shows a circuit using a JFET in a common-gate circuit. The common-gate circuit is analogous to common-base bipolar transistor and grounded-grid vacuum tube circuits. The common-gate circuit, in effect, electrostatically shields the input from the output, thereby eliminating the need for neutralization. This particular circuit uses a series-fed drain circuit tuned by a combination main capacitor and trimmer capacitor. The output signal is coupled to the next stage through a fixed capacitor. The input is broadly tuned by a transformer with a center-tapped primary used to match the balanced 300-ohm antenna lead.

The rf amplifier circuit in Fig. 3-11 uses a dual-gate IGFET as the active element. In this drawing, we show those internal protection diodes mentioned earlier. Because of static buildup on the antenna line and other problems, it is almost mandatory that the rf amplifier transistor be so protected. In the mixer stage, the only critical time for the transistor is during installation or troubleshooting. During that period it is possible for body or tool static charges to arc across the gate insulation and thereby destroy the transistor. Diode protection makes it possible to handle IGFETs without any special care, other than that given any solid-state component.

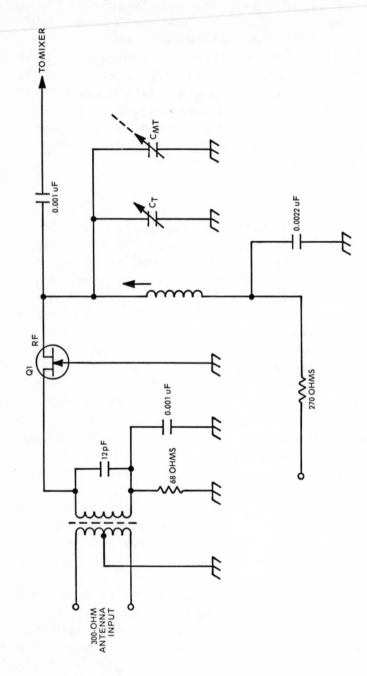

Fig. 3-10. Grounded-gate JFET rf amplifier circuit.

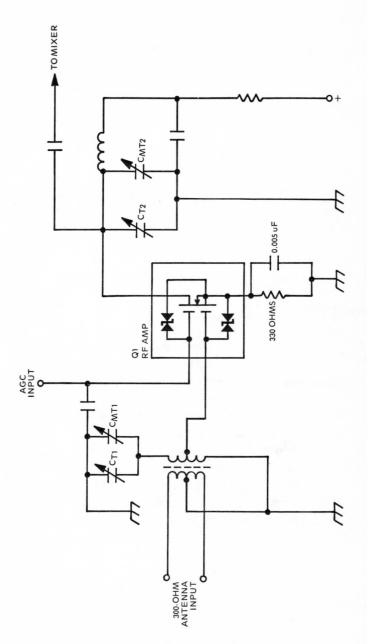

Fig. 3-11. Dual-gate IGFET (MOSFET) rf amplifier circuit.

TO MIXER

AGC
INPUT

Q1
RF AMP

330 OHMS

0.005 uF

C_MT2

C_T2

C_MT1

C_T1

300-OHM
ANTENNA
INPUT

63

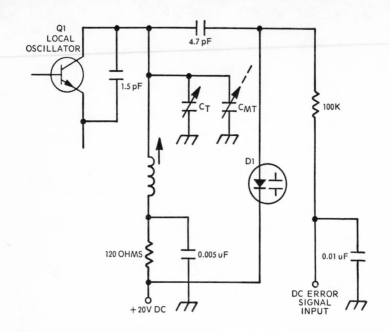

Fig. 3-12. Tank and afc circuit used in a typical local oscillator.

Local Oscillators

One of the original obstacles to widespread introduction of FM in a VHF broadcast band was the necessity of finding a stable local oscillator circuit. In those early days, heat from the vacuum tube circuitry was great enough to cause such a high degree of thermal drift that any hope of stability was abandoned. Now, modern, all solid-state design, effective automatic frequency control (afc) circuits, small high-quality components, and the ability to effectively and inexpensively regulate the power supply output have resulted in oscillators that exhibit exceptional stability characteristics. Regarding those early FM tuners, some technicians said the oscillators were so critical that they would change stations if someone merely walked about the room. Now, the problem of frequency drift has been licked to such a degree that quality FM broadcasting is possible with only a minimum of attention by the user.

The partial circuit of a typical FM local oscillator is shown in Fig. 3-12. This is a variable-frequency oscillator that operates over a range of approximately 98 to 119 MHz. When the 10.7 MHz i-f is subtracted from these figures, we arrive at the frequency range of the U.S. FM broadcast band. In most receivers, the local oscillator is a bipolar transistor, even if the rf amplifier or mixer uses another type of device. Also, in most cases, the oscillator operates in a common-base circuit. Feedback is enhanced by the addition of a small-value capacitor connected between the emitter and collector. In the case of the circuit shown, the external feedback capacitance is 1.5 pF, a value that is almost typical of these circuits. Where the power supply is of the polarity shown in Fig. 3-12, the circuit is almost always series fed. The main tuning capacitor and trimmer capacitor are connected from the collector of the oscillator transistor to ground. This is the usual practice because these capacitors are generally part of a two- or three-section ganged variable capacitor assembly in which the frame is electrically common to all sections. From a mechanical point of view, it is a lot simpler to be able to ground that common frame.

Diode D1 is the afc device. It is a variable capacitance diode (varactor). All pn junctions, whether in diodes or transistors, exhibit a certain amount of capacitance due to internal charges at the depletion zone existing at the junction. In most cases, the diode manufacturer tries to eliminate as much of this capacitance as possible. In the case of the varactor, however, steps are taken in the manufacturing process to enhance this property to a point where the capacitance is usable and predictable. The capacitance varies in value as the reverse bias across the junction is changed. In the circuit in Fig. 3-12, the diode is isolated from the rest of the oscillator tank circuit by a 4.7 pF ceramic capacitor. The function of this capacitor is to block dc while still allowing the varactor to function as part of the resonant tank circuit. The capacitance of the varactor changes with variations of the dc level of the afc error signal derived in the FM detector.

Automatic Frequency Control

The automatic frequency control, or afc, in modern receivers is good enough at its job to effectively eliminate the

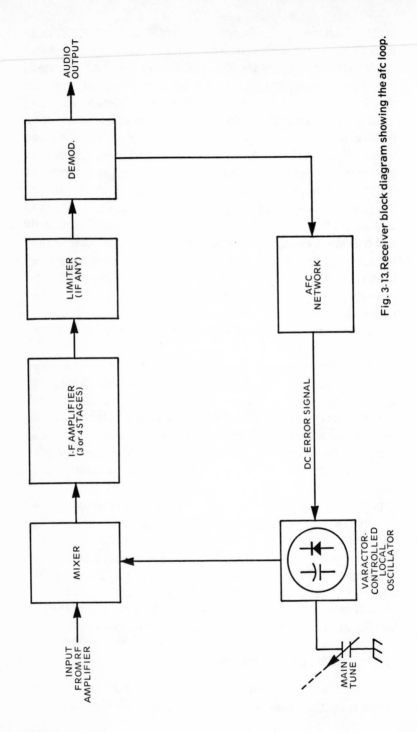

Fig. 3-13. Receiver block diagram showing the afc loop.

drift problems associated with FM in early times. The method of achieving afc is shown in Fig. 3-12, where a varactor is in parallel with the FM local oscillator tank circuit. In Fig. 3-13 you see a block diagram of a typical network. Most FM detectors produce either a specific voltage level or, more often, 0V when the receiver is tuned dead center on an FM signal. If you move off center in one direction, the voltage will either go positive or negative from the quiescent point. The opposite polarity, although the absolute magnitude will be the same, will appear if you change the tuning a like amount in the opposite direction.

The afc network decouples any residual rf components appearing on the dc error voltage and shapes the absolute amplitude to the specific needs of the circuit. The dc error signal causes the capacitance of the varactor to change. When the local oscillator is pulled back on frequency, the error voltage disappears. The net effect, if the afc circuit is well

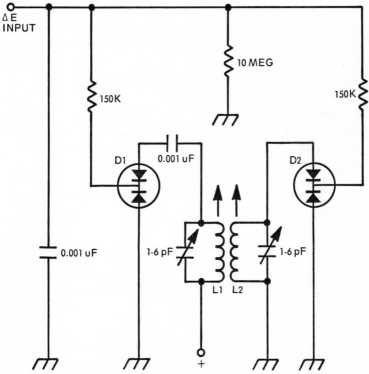

Fig. 3-14. Schematic of a varactor-tuned front end.

designed and working, is to keep the signal centered in the receiver's passband where no error signal is generated.

SPECIAL TUNING CIRCUITS

Varactors are used in a wide range of applications, in addition to oscillator afc. In Fig. 3-14, we see an FM band resonant tank circuit such as might be used to couple the rf amplifier to either the antenna system or the mixer stage. The main tuning element in this circuit is the varactor. The small variable capacitors are used as trimmers to allow both halves of the circuit to track regardless of differences in the absolute characteristics of each individual varactor. Tuning is accomplished by applying a dc reverse bias to the junction terminal of each varactor pair.

These diodes are a little different from the varactors used in the afc circuit in that they are actually two varactor diodes connected back-to-back but housed in a common epoxy case. In most receivers using this system, the dc tuning voltage is derived from a potentiometer ganged to the receiver's main tuning control. In Fisher Radio Company receivers, the pot is mounted as part of the AM band tuning capacitor assembly. A typical varactor-tuned network is shown in Fig. 3-15. A large potentiometer functions as the main tuning control, while a smaller pot serves as a trimmer. The dc voltage supply to this circuit must be extremely well regulated. If there is any

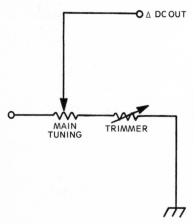

Fig. 3-15. Main tuning and trimmer potentiometers are used to supply a dc voltage to the tuning varactors.

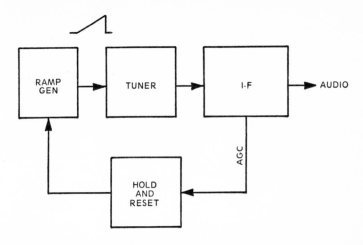

Fig. 3-16. Block diagram of a signal-seeking or autoscan system designed for voltage-tuned receivers.

change in the absolute voltage level of this supply, the frequency of the local oscillator and the rf amplifier will also change. Fisher uses the voltage-tuning concept in their all-electronic "signal seeker" and pushbutton selection features, which are offered on some of their models. Those who are familiar with automobile radios can appreciate some of the difficulties associated with electromechanical signal seeker design. It requires many amperes of current to recock the power solenoid used in many of those systems. Seekers that use motors for mechanical power seem little better off because those motors apparently have a short life expectancy. The use of a voltage-tuned front end makes the all-electronic seeker a reality.

The elementary, and somewhat simplified, block diagram of an electronic signal seeker is shown in Fig. 3-16. A ramp generator is little more than a very-low-frequency sawtooth oscillator. This circuit produces a slowly rising dc level at a frequency considerably less than 1 Hz. The ramp output provides the control voltage input to the tuner front end. The receiver tuning scans the FM broadcast band from the low end to the top end as the ramp rises from zero to maximum. When the ramp reaches its peak value, it suddenly drops back to zero and starts over again. At this point the receiver im-

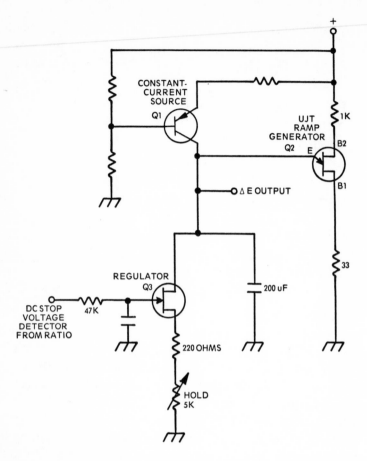

Fig. 3-17. Simplified schematic of the ramp generator and hold regulator circuits used in the Fisher 45OT receiver.

mediately shifts back to the low end of the band in response to the ramp level.

The hold and reset circuit is controlled by a voltage from the i-f circuit. The control voltage halts the further rise of the ramp voltage, yet prevents it from deteriorating, whenever a signal is sensed. If the received station is not to the liking of the operator, he manually reinitiates the seek function, in which case the receiver begins tuning toward the next-higher-frequency station. It would do little good for the receiver to be reset to the low end of the dial at this point. If that were the case, the receiver would only be able to lock on the lowest-frequency station in the area; an admittedly ridiculous situation.

A partial schematic of the ramp generator used in Fisher receivers is shown in Fig. 3-17. Unijunction transistor (UJT) Q2 has a unique characteristic that allows its use as a ramp generator. Under normal circumstances with zero voltage between the emitter and base 1, there is no current (except leakage values) flowing between base 1 and base 2. This situation will hold true even as the emitter-base 1 voltage begins to rise from zero. A certain threshold voltage between these two terminals must be established to trigger the massive current flow. When this is reached, the emitter junction suddenly becomes a low resistance.

The voltage on the emitter is controlled by a bipolar transistor acting as a constant-current source. This allows a relatively linear charging rate for the 200 uF capacitor. The voltage across this capacitor is the voltage that controls the emitter-base 1 junction on the UJT. When it reaches a certain level, the emitter junction breaks over and rapidly reduces the charge on the capacitor to zero. As a result of this action, the voltage function from the emitter to ground is a sawtooth or ramp. It is this ramp function that is used to scan the FM band electronically.

Control over the tuning process and the derivation of the stop function is the job of regulator transistor Q3. When the FM detector senses a station, a stop voltage is applied to the gate input of the JFET, Q3. This causes a current to flow, which is equal to the charge current applied to the capacitor by Q1. Under these circumstances, the voltage drop across the 200 uF capacitor remains at the same level. It will neither rise nor drop as long as the dc bias from the detector is applied to the gate of Q3. When the user wishes to resume scanning the band, the gate voltage is removed by the scanning-mode switch. The capacitor once again begins charging, only this time it starts at the potential it held when the detector caused it to cease scanning. A potentiometer in the source circuit of the JFET allows the service technician to set the current flow through Q3 to the precise value that will keep the capacitor charged at the specified level.

The circuit in Fig 3-18 is used in some receivers to provide a tuning meter. The meter allows the user to tune the receiver to the precise center frequency of the incoming signal. This is necessary if the lowest level of distortion is to be realized.

An alternate type of circuit uses a zero-center dc voltmeter to measure the afc error control line voltage. This circuit is used in receivers where the afc error voltage is zero at the correct tuning point. The meter reads exactly zero when the receiver is properly tuned. On either side of the correct point, the voltage is positive or negative. In the "peaking" type of meter circuit, such as that in Fig. 3-18, we are measuring the strength of the received signal, which is maximum when the receiver is tuned properly. Basically, this circuit is simply a dc voltmeter fed by a rectified and filtered sample of the FM i-f signal voltage. Most receivers employ this style of tuning meter. A few use the zero-center type. In some cases, you may see a model that has both meters. Hopefully, they will both indicate the same correct tuning point.

The logic diagram in Fig. 3-19 represents a mute system used in some FM stereo receivers. Transistor Q1 is a composite amplifier in the multiplex section of the receiver. It could also be one of the preamplifiers or emitter-follower

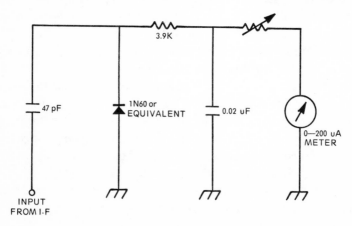

Fig. 3-18. Typical tuning meter circuit.

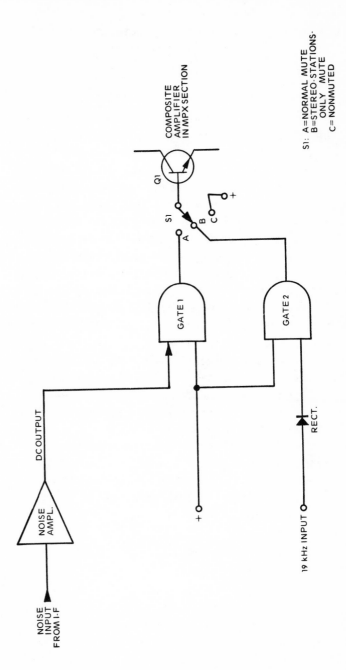

Fig. 3-19. Mute system logic flow chart.

buffers that are often placed between the FM detector and the input of the multiplex decoder. In any case, the muting stage must be at a point where it can easily affect the output of both channels. This means that the controlled stage must be at a point where the two stereo signals have not been separated. Otherwise, there would have to be a separate controlled stage for each channel.

With S1 in position A (Fig. 3-19), the mute opens, thereby allowing the signal to get through to the decoder when the receiver "quiets" during reception of a signal. A noise amplifier takes a sample of the signal from the i-f amplifier strip. This noise signal level is quite high when no signal is present. The output of the noise amplifier is rectified to produce a dc control signal. This is applied to the mute gate input. When a signal is present in the receiver's passband, the dc output is removed and gate one allows the receiver to pass the signal.

With S1 in position B, the receiver will operate only when a stereo station is tuned in. Gate two is turned on and off by the 19 kHz pilot signal transmitted by the FM stereo broadcast station. This signal is rectified to produce a dc level that turns the gate on and allows the signal to pass.

In the last position of switch S1, position C, the transistor in the controlled stage is connected much like a normal amplifier to a source of fixed dc bias voltage. In this position, the receiver behaves as a normal receiver and passes all of the signals in the passband. This includes weak forms of interference not strong enough to trigger the threshold of the mute sections and that abundant hiss for which FM is famous. Most listeners prefer to use the mute function because that hiss is so high that it is considered extremely annoying to those in search of good listening.

Two- and Four-Channel Systems

CHAPTER 4

One of the earliest improvements in electronically reproduced audio was the vast improvement in the frequency response offered by home entertainment systems. This led to the concept of "high fidelity"—a close correlation of the original performance and the reproduced product. No matter how faithful the tonal rendition of the monaural high-fidelity system, there is still an element of the original sound that is missing: directional effects. In a concert hall, you can pinpoint the location of the various instruments and performers because of the bilateral nature of your hearing system. Also to be considered are the effects of reverberation in the hall. Two- and four-channel stereo music systems represent attempts by the audio industry to heighten the illusion of realism offered by recorded and broadcast music. The illusion of separation is accomplished by allowing the directional information to be presented to and reproduced by your home music system.

One of the requirements officially demanded of the FM stereophonic concept was that the resultant signal had to be compatible with existing monaural receivers. In other words, a standard (monaural) FM receiver had to be able to receive a stereo broadcast exactly as if it were in mono. Actually, there were several proposals for stereophonic broadcasting that were never seriously considered because monaural listeners could not receive the broadcasts undistorted.

The compatibility requirement imposed by the Federal Communications Commission (FCC), a federal agency which regulates broadcasting, led to the development of the two-channel multiplex encoding system shown in Fig. 4-1. Left-

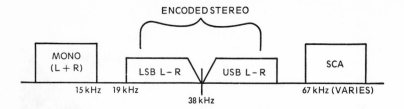

Fig. 4-1. Diagram showing the spectral location of the audio and encoded signals in stereo multiplex.

and right-channel audio signals are mixed together in one stage of the transmitter to form a monaural signal designated "Left plus Right" or simply "L + R." This signal, which is merely a linear mixture of the two channels, is the monaural signal that gives this system the quality of compatibility demanded by the FCC. In order to eventually extract the left- and right-channel signals at the receiver, we need an encoded signal containing the separation information.

The first step in the production of the encoded signal is to invert the phase of the right channel (180 degrees) relative to the same signal fed to the L + R mixer. The inverted signal is designated the "minus R" or "—R" component. It is mixed with the +L signal in another linear mixer to produce a composite known as "L — R." In this case, the minus sign preceding the R simply indicates that it is a phase-inverted copy of the right-channel signal. The L — R signal becomes truly encoded when it is fed to the last stage in the stereo generator. This is a balanced modulator that converts the L — R into a double sideband suppressed carrier (DSSC) signal centered about a carrier (called a "subcarrier" to distinguish it from the FM carrier) frequency of 38 kHz.

The encoded subcarrier signal includes both upper and lower sideband components. Since the audio spectrum allowed FM broadcasters extends to 15 kHz, the subcarrier sidebands extend outward from 38 kHz plus and minus 15 kHz. This means the encoded signal occupies spectrum space between 23 kHz and 53 kHz from the main carrier. A pilot signal at one half the subcarrier frequency (38,000 / 2 equals 19,000) is also transmitted so that an exact replica of the subcarrier can be reconstructed at the receiver.

76

In some localities certain broadcasters are also permitted to transmit "storecast" background music on a second sub-carrier called the "SCA channel." This programing is usually centered at approximately 67 kHz, although some stations are known to use others.

The purpose of the stereo decoder is to take the L + R and L — R (encoded) signals received from the detector and process them into the formerly discrete left- and right-channel signals. This is accomplished by allowing the two components of the composite to mix together in a bilateral product detector that is switched on and off by a signal that has the same frequency and relative phase as the original subcarrier at the transmitter.

The concept of product detection implies multiplication. In this case, the output of a product detector is a "product" of the two input signals. The classic, simplified expressions for the derivation of the two discrete signals is:

$$(L + R) \ (L - R) = 2L - RL + RL - 2R \ = 2L - 2R$$

The action of the product detector produces two discrete signals representing the left and right channels of the original stereo signal.

The block diagram in Fig. 4-2 identifies the stage functions in a typical (almost classic) two-channel stereo FM receiver. The composite signal containing the L + R, L — R (encoded) and 19 kHz pilot information is taken from the output of the FM detector circuit. These signals are first fed to a preamplifier that functions as a signal splitter or signal "router." From one output of the router, we extract the 19 kHz pilot signal. This signal is amplified and then fed to a frequency doubler, where it is converted to 38 kHz, the original subcarrier frequency. The other output from the router contains the L + R and L — R (encoded) signals.

These may or may not receive additional amplification after leaving the router. Both the L + R—L — R (encoded) composite signal and the regenerated subcarrier are fed to the

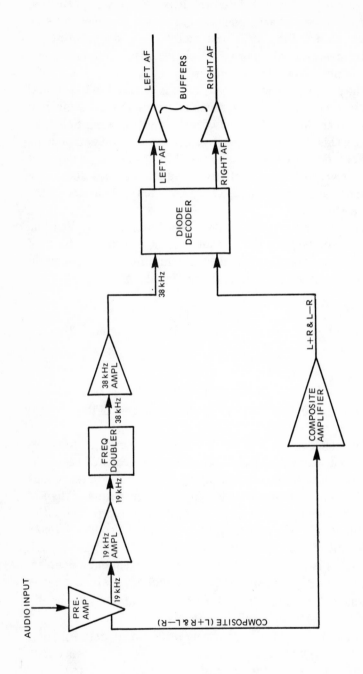

Fig. 4-2. Block diagram of a transistor stereo decoder.

diode product detector decoder matrix. From the decoder, the now discrete left- and right-channel audio signals are fed first to 38 kHz filters and then to buffers and the rest of the audio amplification chain.

SIGNAL SPLITTER / PREAMPLIFIER CIRCUITS

In order to gain an appreciation of how a typical two-channel stereo decoder section operates, we consider each stage individually. As stated earlier, the first stage in most decoders is the splitter / preamplifier.

An example of this type of circuit is shown in Fig. 4-3. The composite signal from the detector first encounters a parallel resonant LC trap circuit tuned to the 65-72 kHz SCA band. This trap eliminates the SCA signal so that it cannot interfere with the received stereo signal. The collector of the preamplifier is connected to the primary of a resonant tank transformer tuned to 19 kHz. This picks off the pilot signal so that it can be used to regenerate the subcarrier. The emitter circuitry usually contains a tank tuned to 19 kHz, which is a trap to prevent the 19 kHz signal from entering the composite line. The L + R and L — R (encoded) signals, less the pilot and any SCA signals that may have been present, are picked off the emitter of the splitter preamplifier stage. These signals, known as the composite, are fed through either an RC network or a composite amplifier to the diode product detector matrix.

SUBCARRIER REGENERATION

There are actually several ways to regenerate the sub-carrier so that the stereo information can be recovered. In some early receivers, a phase-locked oscillator was used. In that system, a 38 kHz oscillator was phase- and, of course, frequency-locked by a synchronizing signal derived from the 19 kHz pilot amplifier. The locked oscillator faded from popularity because the frequency doubler (see Fig. 4-4) does as good a job and is a lot easier to keep on frequency. There is little, if any, detectable difference between the two circuits on the output end. They are both able to deliver the correct

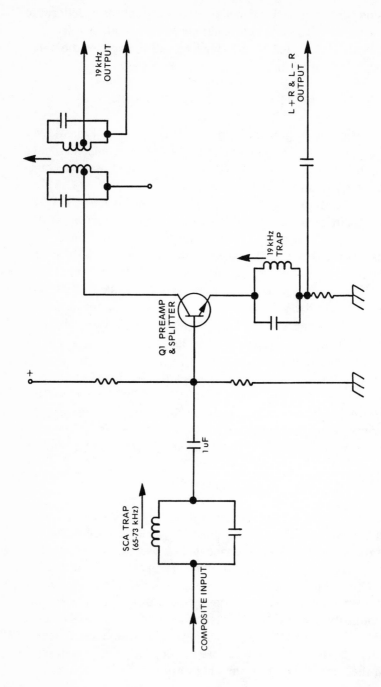

Fig. 4-3. Simplified schematic of a multiplex preamplifier and splitter stage.

regenerated subcarrier signal. The locked oscillator, however, tends to jump out of lock under the right circumstances. The frequency doubler circuit, on the other hand, will operate satisfactorily as long as there is a 19 kHz signal present with sufficient amplitude to forward bias the diodes.

It is worth noting that, recently, there has been a return to the locked oscillator concept. Certain modern integrated circuit stereo decoders use a 76 kHz locked oscillator and J-K flip-flop frequency dividers in a phase-locked-loop (PLL) system.

The secondary of the 38 kHz frequency-doubler transformer is like a full-wave power supply transformer. It is center-tapped to provide equal but opposite polarity signals to each of the diodes (Fig. 4-4). The rectified output, as in power supply circuits, is a train of half-wave pulses. The stereo decoder, unfortunately, cannot use this pulsating dc (with 38 kHz ripple) output. It requires a sine-wave with the same phase and frequency as that which originally existed in the transmitter. The pulsating dc is converted to a 38 kHz sine wave by the "flywheel effect" of the 38 kHz tuned transformers in the subcarrier amplifier circuit.

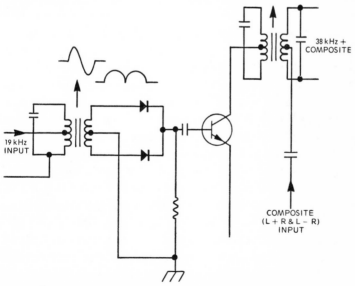

Fig. 4-4. Simplified schematic of a full-wave frequency doubler used to convert the 19 kHz pilot tone to the 38 kHz subcarrier.

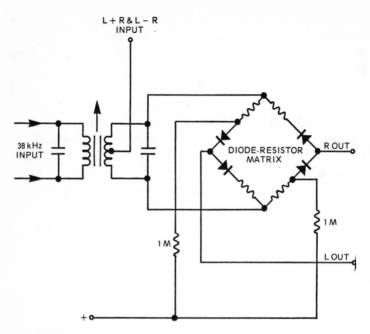

Fig. 4-5. Resistor-diode matrix decoder used in early Fisher receivers.

DECODER CIRCUITS

By far, the most popular form of product detector for decoding the left- and right-channel signals is the circuit shown in Fig. 4-5. It is called the "balanced-bridge" or "ring" demodulator. It is not unfamiliar to those areas of electronics which are concerned with sideband and other forms of suppressed-carrier communications. The encoded, or "multiplexed" signal used in FM stereo broadcasting is actually a double-sideband suppressed carrier form of signal.

In almost every case, the composite signal, consisting of L + R and L — R (encoded), is applied to the product detector through a center tap on the secondary of the 38 kHz output transformer. The adjustment of this transformer and the degree of balance between the diodes has a profound effect on the amount of separation realized by any specific receiver. Under ideal circumstances (rarely attained in real life), if one diode becomes defective, all four should be replaced with a set matched for this purpose by the manufacturer.

The output of this circuit usually passes through an RC deemphasis network designed to restore the frequency response relationships lost at the transmitter and to suppress any residual 38 kHz signal remaining after detection. These filters are often packaged in one of several styles of ceramic modules offered by various companies.

The circuit in Fig. 4-6 is a refined version of the diode-resistor matrix in Fig. 4-5. In this configuration, there are actually two ring demodulators, one for each of the two channels. It is basically a full-wave version of the circuit in Fig. 4-5. This circuit may prove popular in the future because of certain advances in IC technology. Already, companies in the semiconductor field offer IC packages containing four or eight diodes. The buyer has the option of purchasing chips with the diodes left "floating" or already preconnected into the diode ring network.

This IC approach to diode fabrication has certain advantages. For one thing, the diodes are better matched, since they all come from the same batch of molten semiconductor material at the same time and at the same temperature. The

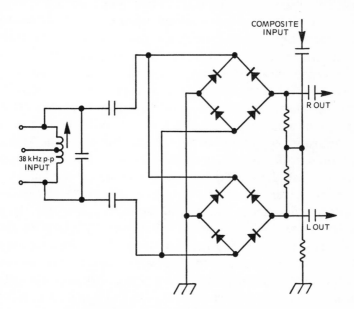

Fig. 4-6. Dual matrix diode decoder circuit.

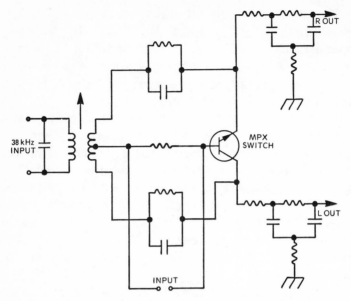

Fig. 4-7. Low-cost single-transistor switch decoder circuit.

absolute specifications will vary from chip to chip, of course, but the diodes will remain relatively well matched to each other. It is also true that they will remain matched because they share a common substrate and package. This keeps them in the same thermal and contaminant environment.

In case you are wondering how each ring knows whether it is for the right or left channel, please keep in mind that it is the phasing of the subcarrier signal that sets this parameter. Incorrect phasing of the subcarrier will result in a deterioration of separation between the two channels. This is the reason that it is important to leave stereo adjustments alone unless the correct alignment equipment is available.

For general interest, examine the single transistor switching decoder in Fig. 4-7. In this circuit, a bipolar transistor is switched into and out of conduction by the 38 kHz subcarrier, while composite information is applied to the base. The result is a decoder with reasonable stereo separation at a low cost to the manufacturer. This type of circuit is generally found only in certain console receivers and in low- to medium-priced bookshelf equipment.

The price of a stereo receiver, by the way, often has little to do with the type of circuits used in this section of the system. This was pointed out to the author when a comparison between an "under 90 dollars" table model stereo radio and an "over 400 dollars" receiver showed that they both used the same circuitry in the stereo decoder. In fact, the similarity between the two schematics was so great that it looked like the same man drew both diagrams!

IC STEREO DECODERS

The average stereo decoder section block diagram is essentially the same from one receiver to another. The specifics vary only slightly from one to the other. Since this has been the case for many years, it was only natural that some company would eventually produce an integrated circuit to decode stereo signals in FM receivers. Motorola, in conjunction with a stereo equipment manufacturer, first broke the ice with the earliest of their MC1300 series ICs. This line includes the MC1304P, MC1305P, and MC1307P. Although the line has been expanded to the MC1310P, a PLL design, those earlier types were the first, and they are used in great abundance. Today, almost every semiconductor manufacturer supplying the consumer electronics field offers some type of IC stereo decoder.

The three Motorola chips listed differ only in specific details. The overall block diagram is pretty much the same in basic concept for all three units. The block diagram for the MC1304P is shown in Fig. 4-8. The first stage encountered by the composite input signal is a preamplifier. If the signal is monaural FM, or if it is an audio signal derived from another source such as the AM radio section, it will lack the pilot signal. In the absence of the 19 kHz pilot, the IC merely splits the signal into two identical parts and passes it along to the two outputs without further processing. This greatly simplifies bandswitching and signal mixing problems. Except for some amplification and the splitting, there is no action on the signal inside the IC unless the presence of the 19 kHz pilot is detected.

The pilot is sensed by the 19 kHz amplifier and doubled to 38 kHz in a tank circuit. The output of the 38 kHz tank is fed

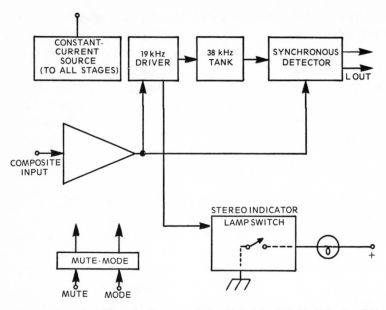

Fig. 4-8. Functional block diagram of the Motorola MC1304P Series FM multiplex stereo decoder IC.

directly to the synchronous transistor product detector. The other input to the synchronous detector is the composite output from the preamplifier. One other output from the 19 kHz circuit is the drive signal to the stereo lamp-switching transistors. These transistors are used to turn on a lamp that lets the user know when he is tuned to a stereo station. They are connected to act as an spst switch to ground. The switch is used to ground one side of a stereo indicator lamp or the base resistor of an external lamp-switching transistor if the current requirements of the indicator lamp exceed 40 mA.

The last section of the decoder IC has two extra control inputs that provide a "mute" function and a "stereo/mono" function. When the "mute" input is grounded, the output of the receiver is silenced. When a dc voltage of +1.5V is present at this terminal, the output is allowed to pass a signal. This dc voltage can be developed by rectifying a sample of the i-f signal, so that it exists only when a station is tuned in.

The mono/stereo switch uses a +1.5V dc level to turn on the stereo function. The control voltage can be derived either from a panel switch or from an automatic circuit that senses a condition where the signal level is too low for decent decoding of the stereo information. Under these conditions, it is possible to receive good monaural material from a station that will produce only marginal and very noisy stereo programs.

The external circuitry used with the Motorola MC1304P IC stereo decoder is shown in Fig. 4-9. In most sets using this IC decoder, there are three tank circuits: one tuned to 38 kHz and two tuned to 19 kHz. Shunting the output lines to the positive side of the power supply are the deemphasis networks. FM broadcasters preemphasize the high audio frequencies prior to transmission in order to improve the signal-to-noise ratio at the receiver. This must be counteracted at the receiver end or the high-frequency components in the recovered audio will be too high in amplitude. This results in a sound that is often described as "harsh" or "overly bright."

The deemphasis network does not affect the AM audio fed through the chip because it acts only on those frequencies above 5 kHz. These are above the normal passband allowed to AM broadcasters. In applications where neither mute nor

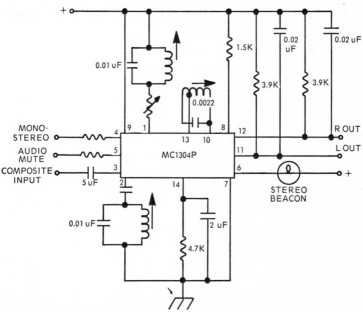

Fig. 4-9. Schematic of the external circuitry used with the Motorola MC1304P.

mono/stereo functions are desired, the appropriate control inputs are returned to a positive dc voltage source via a fixed resistor. This permanently biases that function in the stage. These ICs will operate over a supply voltage of 8 to 20V.

A newer IC decoding approach takes us back to the use of a phase-locked local oscillator. This is illustrated in Fig. 4-10. This type of circuit overcomes the lock instability problems of earlier attempts to use an oscillator through a technique called the "digital phase-locked loop" (PLL). With this technique, a feedback control loop keeps the local oscillator locked to the appropriate frequency. The first stage in the processing of the stereo signal is once again a type of splitter. In this application, the splitter divides the signal into two quadrature components. These are fed in parallel to the matrix decoder. Also, they are simultaneously fed to a phase-lock detector, a 90 degree (quadrature) phase detector, and an L — R detector. The output of the primary phase detector is fed to a local oscillator control circuit. This drives a voltage-controlled oscillator (VCO) running at approximately twice the subcarrier frequency (2 x 38 kHz equals 76 kHz). The VCO produces an exact 76 kHz signal when it is properly phase locked to the 19 kHz pilot tone.

The beauty of this system is that only a rough adjustment of the VCO LC tank circuit is necessary for the circuit to operate. In older locked oscillator circuits, the adjustment of the oscillator tank was hypercritical; you needed an elaborate test setup. In this circuit, only a stable and accurate signal source is required. When the tank is properly adjusted, the stereo lamp will turn on, which indicates correct adjustment. Some PLL decoders, notably the MC1310P by Motorola, use only a trimmer potentiometer resistor to adjust the oscillator. In these cases, the VCO is an RC type. If you manage to adjust the VCO to within a small percentage of the correct point, the PLL function will take over and do the rest.

The output frequency of the VCO is first divided by a factor of two in a J-K flip-flop stage in order to derive a 38 kHz signal for L — R detection. The output of the VCO also feeds two more J-K stages connected to provide quadrature 19 kHz signals for the phase-lock and quadrature comparator cir-

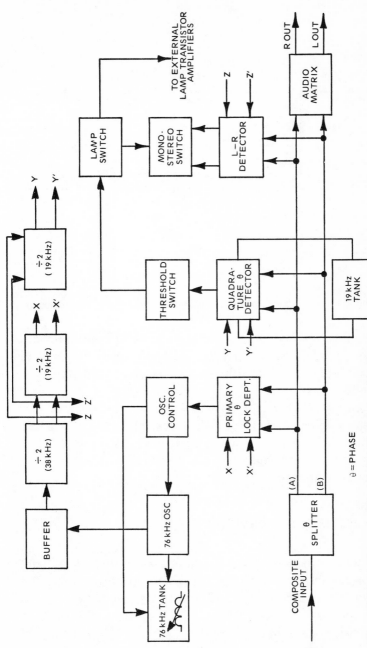

Fig. 4-10. Block diagram of the digital PLL stereo decoder.

θ = PHASE

89

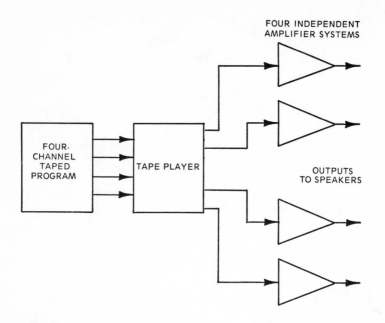

Fig. 4-11. Block diagram of the discrete 4 x 4 x 4 four-channel system.

cuits. A lamp-switching transistor, external to the IC, provides a stereo indicator function. Decoded left- and right-channel audio signals are fed to the rest of the radio through buffer amplifiers.

PLL seems to be the wave of the future for FM receivers. It is a technique that can be used to control the 19 kHz∕38 kHz portions of the stereo decoder, demodulate the i-f signal, and even control the frequency of the VHF local oscillator used in the tuner front end.

FOUR-CHANNEL TECHNIQUES

Since two-channel stereo offers a great deal of realism, four-channel stereo seems, at first glance, to be something of a redundant frill designed mostly to bolster the sales of audio equipment. There can be no doubt, however, that a four-channel system offers "something extra" in the way of realism. Those who have invested in the equipment, and their number grows daily, will attest that "something" is well worth the extra cost.

Basic Quadraphonic Configurations

There are three basic methods for handling a four-channel stereo system. The first of these, detailed in Fig. 4-11, is the 4-4-4, or "totally discrete" system. In this system, the playback equipment is presented with four separate and independent tracks of audio material. No encoding is used. The use of this method is almost exclusively limited to tape, since it is difficult to provide the four tracks necessary on any other medium. RCA first offered four-channel program material in the form of "Quad-8" eight-track tape cartridges. Motorola offered an eight-track "Quad-8" player for automotive installation, while Toyo offered a home unit. In the "Quad-8" format, there are two four-channel programs; each requires four tracks on the tape.

In FM broadcasting and on phonograph discs, we are limited to the use of two channels—because of current regulations on FM, and on discs because of the cutting techniques. In order to bring any four-channel stereo system to its full potential, it is necessary that the broadcasting and disc media be included.

There are two main approaches to the problem. Both approaches, the various matrixes and the RCA-JVC "discrete," encode the signals needed to reproduce the rear channels separate from the normal left and right front channels. Such systems (see the block diagram in Fig. 4-12) are generally referred to as "4-2-2" quadraphonic systems. In this case, the four individual channels of sound are reduced to only two channels by one of several proposed methods. The signals in these channels are described as "left total" and "right total" and are completely compatible with existing two-channel stereophonic systems. The L_T and R_T signals can be used to cut grooves into the surface of a phonograph record or they may be used to modulate an FM stereo transmitter as "$L_T + R_T$" and "$L_T - R_T$" signals.

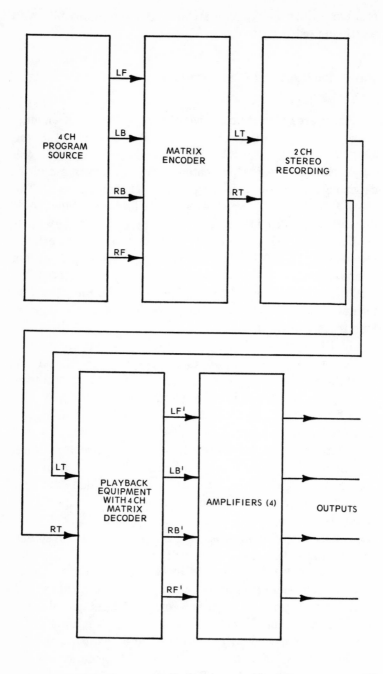

Fig. 4-12. Block diagram of the 4-2-4 matrix four-channel stereo system.

2-2-4 Four-Channel Stereophonic Systems

One of the earliest commercially produced four-channel stereo systems is the "2-2-4" system shown in Fig. 4-13. This is not really a true four-channel system; rather, it is an "ambient sounds" recovery method. In this system, popularized by the Dynaco Company under the "DynaQuad" trademark, four channels of information are recovered from a two-channel stereo source through the use of some unusual speaker connections. DynaQuad takes advantage of the fact that certain directional components due to recording hall reverberation are contained in most present two-channel stereo recordings. This effect is best appreciated when listening to recordings made in concert halls or other large, acoustically "live" rooms.

One of the early 2-2-4 four-channel stereo arrangements is shown in block diagram form in Fig. 4-14. In this system, the left and right front speakers receive the normal left- and right-channel signals. A third channel receives both left and right channels so that it can combine

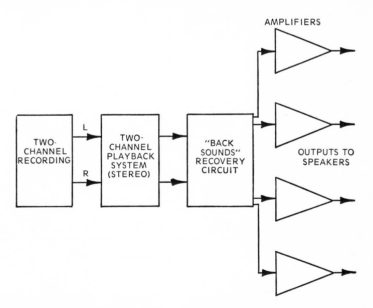

Fig. 4-13. Block diagram of the 2-2-4 "ambient sounds" recovery system popularized by Dynaco.

signals to produce an L — R signal. This composite is fed to a pair of rear speakers in quadrature (90 degrees out of phase). The interconnections should be made inside a stereo receiver. It is the method used by many of the early "internal synthesizer" receivers. Other systems, such as the DynaQuad system mentioned earlier, use speaker connections to achieve the goal of ambience recovery.

The "DynaQuad" label can actually be applied to two different methods of connecting four speakers to a regular two-channel stereo system. In the circuit in Fig. 4-15, we see one of these methods. The left and right speakers are connected with their positive terminals (polarity here refers to an arbitrarily defined polarity used for purposes of speaker phasing) to the amplifier as in normal two-channel stereo practice. The negative terminals, however, are not returned directly to the amplifier common terminals. Instead, they are connected to the positive terminal of a third speaker referred to as the "front speaker." The negative terminal of the front

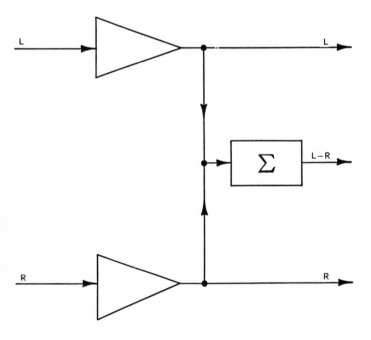

Fig. 4-14. Block diagram of one of the early 2-2-4 four-channel systems.

speaker goes to the negative terminals on the amplifier. This method is reminiscent of the "hole in the center" system of an earlier era. A fourth speaker is placed at a somewhat greater distance from the listener than are the other three. This speaker is connected across the hot, or positive, terminals of the two channels so that it, in effect, bridges the two outputs.

A modification of the basic DynaQuad scheme is shown in Fig. 4-16. In this system the left- and right-channel speakers are connected across their respective amplifier outputs in the normal manner. A second pair of speakers is connected so that their positive terminals go to the hot side of each amplifier output while the negative terminals are jointly returned to common through a resistive pad. The pad has the effect of reducing the output from each of these "rear" speakers. In some versions of this basic system, especially those operated from low-powered amplifiers, there may be a T or L pad attenuator connected to each speaker. This allows the user to adjust what has been described as the "sound focus" of the

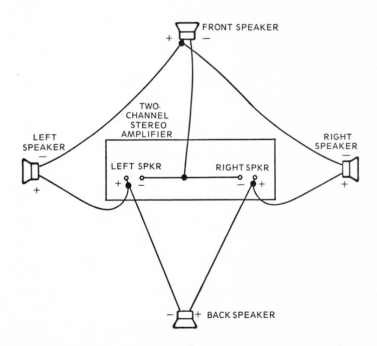

Fig. 4-15. Connection diagram for the basic "DynaQuad" system.

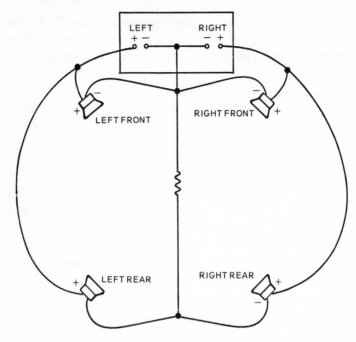

Fig. 4-16. Variation of the "DynaQuad" four-channel technique.

system by custom tailoring the amplitude of the signal fed to each individual speaker.

The DynaQuad system has gained a measure of popularity because all that it requires is one additional pair of speakers and a low-cost control box to insure correct connection. It, and close derivatives, have also been used to simulate four-channel ambience in automotive stereo installations.

Matrix Quad

There are two competing systems trying to gain supremacy in the "true" four-channel phonograph and broadcasting markets. These systems are being pushed by two giants in the recording industry: RCA and Columbia. RCA, in conjunction with JVC of Japan, is pushing their "Compatible Discrete Four" (CD-4), while Columbia is pushing a matrix system developed by their affiliated CBS Laboratories. To make the scene even more confusing, several other manufacturers, include Sansui in Japan and

Electro-Voice in the United States, are offering their own versions of the matrix form of quadraphonic stereo.

On first pass, the various competing matrix quad systems seem to be not a bit compatible with each other. When reduced to arithmetic, however, certain close similarities become apparent. In fact, many systems are so close that many manufacturers are able to offer matrix quadraphonic receivers with a bank of switches to select one of the currently used matrix systems. One manufacturer, Electro-Voice, is offering a universal integrated circuit decoder that is reputed to decode all matrix systems without the need for switching. Such a system greatly simplifies the problem of designing and producing a universal quad receiver. The universal IC decoder is designated "EVX-44" or "New E-V." The older, original E-V system, dubbed the "Stereo-4" is described presently.

The CBS matrix, called the "SQ" system, is shown in Fig. 4-17. The block diagram in this illustration is suitable for most of the matrix systems being discussed at length. In the matrix system, the four channels of audio are reduced to two composite channels in an encoder. The outputs from the encoder

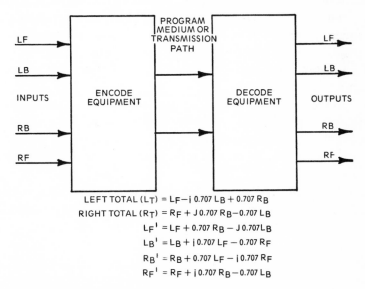

$$\text{LEFT TOTAL } (L_T) = L_F - j\,0.707\,L_B + 0.707\,R_B$$
$$\text{RIGHT TOTAL } (R_T) = R_F + J\,0.707\,R_B - 0.707\,L_B$$
$$L_F{}' = L_F + 0.707\,R_B - J\,0.707\,L_B$$
$$L_B{}' = L_B + j\,0.707\,L_F - 0.707\,R_F$$
$$R_B{}' = R_B + 0.707\,L_F - j\,0.707\,R_F$$
$$R_F{}' = R_F + j\,0.707\,R_B - 0.707\,L_B$$

Fig. 4-17. Block diagram of the CBS SQ matrix four-channel system.

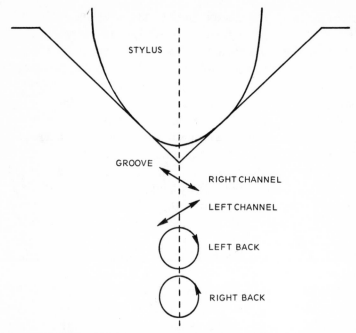

STYLUS

GROOVE

RIGHT CHANNEL

LEFT CHANNEL

LEFT BACK

RIGHT BACK

Fig. 4-18. With the CBS SQ system, the four channels are cut into a single groove as shown.

resemble normal stereophonic signals and can be played on normal two-channel equipment with no apparent deterioration in acoustic quality. Many two-channel listeners even report an increase in the subjective quality called "ambient brilliance" when listening to four-channel sound played on two-channel equipment.

At the receiver end, the two totalized signals (L_T and R_T) are fed to a decoder that recovers the four independent channels of audio present in the original recording. These signals are then fed to four separate amplifier and speaker systems.

The arithmetic equations appropriate to the SQ matrix system are shown on Fig. 4-17. A general equation that might satisfy all matrix systems might look like:

$$L_T \quad L_F + L_R - dR_F - dR_R$$

$$R_T \quad R_F + R_R - dL_F - dL_R$$

These are the equations that could represent the output of a typical matrix encoder. One of the main points of difference among the various matrix systems is the polarity of any particular term and the value of the term "d."

The real innovation of the SQ system is the method used for impressing the information into the grooves of the record disc. In normal stereo recording, the left- and right-channel components are cut into the left and right walls of the groove. The two walls each lie in a plane 45 degrees from horizontal and 90 degrees from each other. In the SQ record, they retain the stereo 45-45 groove system so that compatibility is maintained. In addition to the regular cut, however, they add a new double helical cut (see Fig. 4-18). Rotational groove modulation in the clockwise direction contains the left-back signals, while rotational modulation in the counterclockwise direction contains the right-back sounds.

The Sansui "QS" system is a variation on the matrix theme that is worthy of note. The basic block diagram of this method is shown in Fig. 4-19. The right- and left-front sounds are treated individually in the normal stereophonic manner. The right-rear and left-rear sounds, however, are altered. The left-rear sounds are given a positive phase shift of 90 degrees, while the right-rear sounds are given an equal but opposite negative phase shift of 90 degrees. The phase-shifted LR signal is then combined in a linear mixer circuit with the LF signal. Likewise, the phase-shifted RR signal is combined with the RF signal. The outputs from the two linear mixers form the two totalized signals designated L_T and R_T.

At the decoder, which is found inside the amplifier or receiver in some cases and as an external add-on unit in others, the signals are split up in such a manner that the RR and LR signals are given 90 degree phase shifts in the direction opposite to that which they received in the encoder. This action restores the rear-channel signals to the phase relationships which they had in the original four-channel master recording (actually, the true master probably had up to 16 discrete tracks, but these are mixed down to only four discrete tracks in the disc-making process). Most QS system units include adjustment potentiometers for balancing the relative amplitudes of the different signals.

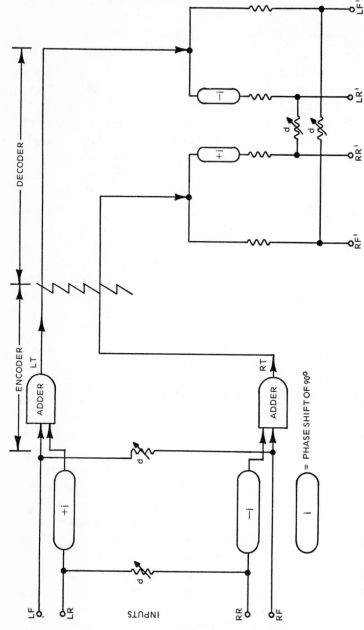

Fig. 4-19. Block diagram of the Sansui QS four-channel system.

Two approaches to the matrix decoder circuit are shown in Figs. 4-20 and 4-21. In Fig. 4-20, we see the original Electro-Voice Stereo-4 decoder, while Fig. 4-21 is a 2 + 2 "logic decoder" circuit used by Fisher in some of their receivers. Notice the multivibrator-like configuration of the latter circuit. These two decoders represent two of several possible approaches to the problem. Unfortunately, neither allows the efficient decoding of all currently contending matrix type quadraphonic systems.

Another, and more viable approach, is the EVX-44 "universal" decoder (Fig. 4-22) by Electro-Voice. This decoder is reputed to offer superior performance when compared to certain earlier semiuniversal types. It is supposed to be especially suited to overcome separation problems in the rear channels caused by a "center soloist" (both L and R front with same signal) situation.

RCA / JVC Discrete (CD-4)

Although called "discrete" by its originators, the RCA/JVC CD-4 quadraphonic system is actually a form of

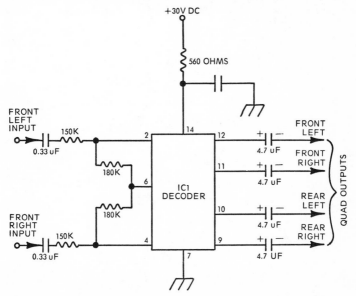

Fig. 4-20.Schematic of the original Electro-Voice Stereo-4 decoder.

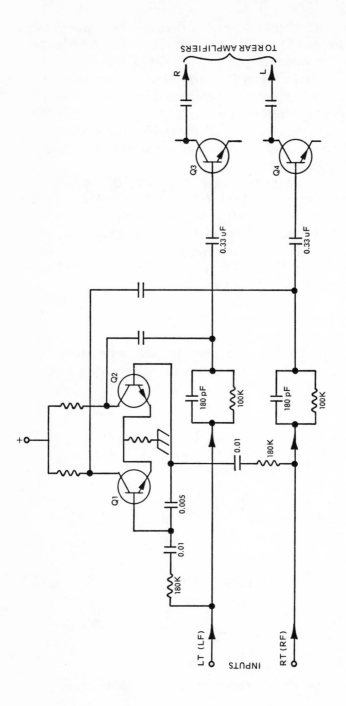

Fig. 4-21. Schematic of the Fisher 2+2 logic decoder.

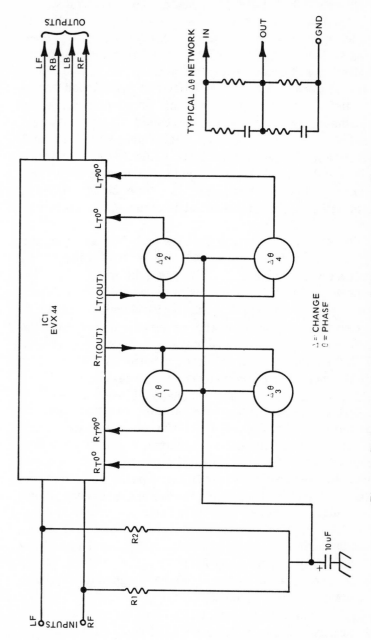

Fig. 4-22. Schematic of the Electro-Voice EVX-4 "universal" decoder.

103

multiplex that is not really unlike the normal two-channel system used in FM stereo broadcasting. To be really discrete, the CD-4, or any, system would have to have four independent channels of audio with no mixing or encoding necessary. Such systems have actually been tried experimentally by cooperating FM broadcasting stations.

In California, FM station KIOI has been broadcasting discrete four-channel signals for some time with remarkable performance results. The system, called "Quadracasting," was developed by Lou Dorren, the engineer credited with perfecting RCA's CD-4 demodulator. While Dorren's approach appears viable, the FCC has not yet sanctioned its use except on an experimental basis. For the time being, at least, "discrete" four-channel material is restricted to tapes and discs.

The method of recording the CD-4 four-channel signal in a single groove is shown in Fig. 4-23. The left wall is modulated by two signals representing a sum and a difference of certain components. The difference signal is phase-modulated about a carrier frequency of 30 kHz. It has positive deviation of 15 kHz and negative deviation of 10 kHz. In the audible portion of the spectrum, we see a sum signal made of channel 1 + channel 2. This is the regular signal required for normal stereo listening (left channel). The difference signal used to phase-modulate the 30 kHz carrier is comprised of channel 1 — channel 2.

On the right wall of the record groove, we see pretty much the same situation. The only real difference is that the signals are comprised of channel 3 + channel 4 and channel 3 — channel 4. As in the case of the left wall, the difference signal is used to phase-modulate the 30 kHz carrier. The "sum and difference" method of encoding the two extra channels allows a CD-4 disc to be played on existing stereophonic equipment. On CD-4 equipment, such discs produce four completely discrete channesl of audio. The playback system must have four of everything normally found in audio systems, plus a decoder and a special pickup cartridge designed to play those ultrasonic carrier frequencies.

A block diagram of a typical CD-4 type encoder is shown in Fig. 4-24. To simplify matters, we have designated channel 1 as "A," channel 2 as "B," channel 3 as "C," and channel 4 as

"D." In the encoder, we see the four mixers necessary to produce the four components of the two different composites. In this case, the B input to mixer 2 and the D input to mixer 3 are first fed through inverter circuits. The purpose of these inverters is to reverse the phase of the input signal. This means the output signal will be 180 degrees out of phase with the input.

In the arithmetical shorthand adapted for this discussion, the inverter causes B to become −B and D to become −D. The four mixers produce the following outputs: (1) A + B, (2) A − B, (3) C − D, and (4) C + D. The outputs of the mixers 2 and 3 are fed to separate phase modulators. These stages use the C − D and A − B signals to phase-modulate the 30 kHz ultrasonic carriers. Mixers 5 and 6 form the two totalized

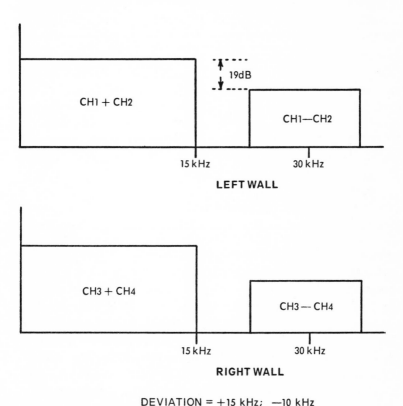

DEVIATION = +15 kHz; −10 kHz

Fig. 4-23. The RCA CD-4 disc recording technique utilizes ultrasonic "carriers."

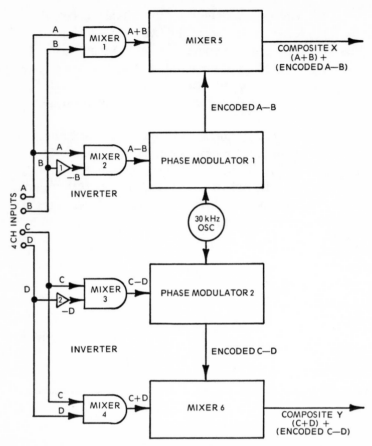

Fig. 4-24. Block diagram of the RCA CD-4 encoder.

signals actually fed to the disc-cutting equipment. Mixer 5 receives the phase-modulated 30 kHz (A — B) at one input and the A + B at the other input. The output of mixer 5 is a composite of (A + B) + (A — B phase-modulated on 30 kHz). Mixer 6 does this same trick with the C + D and C — D signals.

At the playback end, the input signals are separated by filtering. The A + B and C + D are split from the composites by low-pass filter networks. The signals modulating the 30 kHz carriers are separated by bandpass filters. The original A — B and C — D signals are recovered through the use of an FM detector type of circuit operating at 30 kHz. The actual job of recovering the four channels of audio is the function of a special four-input product detector network. From inputs representing $A + B$, $A - B$, $C + D$, and $C - D$, we acquire the original outputs of A, B, C, and D.

One of the most interesting facets of any struggle for system supremacy is the refinement of the basic process that inevitably leads to a superior product. Initially, there was competition over which TV system to use, followed by competing color TV systems prior to adoption of the NTSC standards. In the early days of stereo, there was keen rivalry over various disc systems, recording methods, and broadcasting techniques. In the end, in each case, the system which was the most compatible with existing systems won out. In other words, before our present system for two-channel stereo was accepted, the proponents had to demonstrate that monaural listeners would not experience any deterioration in the quality of the program material received. In actual practice, most users received superior programing and audio from the stereo stations because of the more acute demands of stereo buffs.

In the present quadraphonic war, the test of survivability might easily be the customer's appraisal of best compatibility. The Dorren system (a matrix-type quadraphonic) is already being tested by some stations. It is felt by some experts that this system is already legal under existing FCC rules and regulations. When the industry finishes making war and begins to take seriously the business of selling records, we will eliminate the losers and concentrate on the winners.

CHAPTER

5

AM Receiver Circuits

The AM broadcast band is of slight interest to most FM stereo receiver owners. The high noise level, coupled with the "low-fi" 5 kHz audio passband permitted by the FCC, makes AM considerably less desirable than FM. In most stereo receivers incorporating an AM section, the circuitry may tend to be a little less involved than the quality of the rest of the set may tend to indicate. Only a few manufacturers put any amount of real effort into the production of an AM section for their line of stereo FM receivers. If the user lives in one of the more remote sections of the country, he may well need an AM band. For most of the population, however, the AM band is offered merely as an extra feature so some salesmen can yell, "Lookie here, before you buy!"

In Fig. 5-1, you'll see the block diagram of a typical superheterodyne radio receiver. Most of the radios manufactured since the early 1930s (and many before that time) use the superhet technique because it is inherently easier to design, operate, and is more stable than regenerative or trf type receivers.

In some designs, the signal picked up by the antenna is fed to a stage called the rf amplifier. From there, the strengthened signal is passed to either a converter or mixer stage. In the mixer, the rf signal heterodynes with another rf signal generated by a local oscillator. The result is a difference-frequency signal. In the converter, we see the same action, except that a single transistor or IC performs both mixer and local-oscillator functions. The difference produced by the mixing action is a third frequency called the intermediate frequency (or simply i-f). It is in the i-f amplifier chain that the receiver offers most of its gain and selectivity, since fixed-frequency amplifiers are much easier to design and control than are variable-frequency stages such as the rf amplifier.

Following the i-f amplifier is an envelope detector designed to recover the audio program material modulating the rf carrier signal. This stage may feed an extra audio preamplifier, but it is generally used to drive the control amplifier directly.

AM ANTENNAS

A length-tuned outdoor antenna for the AM band would have to be on the order of several hundred feet long. Because of this, all home antennas for this frequency range are compensation types. The most popular of these is the "loopstick" antenna. One example of this style antenna is shown in Fig. 5-2. The antenna consists of several windings of fine wire on a ferrite form. One winding is simply a low-impedance coupling to match the base input of either the rf amplifier or a converter. In cases where there is no rf amplifier, there often is a third winding on the loopstick to augment the local oscillator feedback path. In the example shown, however, the loopstick

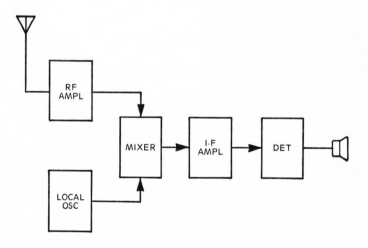

Fig. 5-1. Block diagram of a typical AM superheterodyne receiver.

is designed to feed an rf amplifier stage. The other winding on the antenna is the tuned inductance in the input tank circuit.

On many receivers the loopstick is external to the rest of the receiver, usually mounted to a swivel assembly that can be positioned for best reception. This is a decided advantage because a loopstick is highly directional. It is so directional, in fact, that many marine band receivers use a loopstick as the sensing element in the radio direction finder. The mechanical design of the typical loopstick makes it appear extremely attractive as a handle for picking up the receiver. The author's advice on this score is identical to the manufacturers: don't.

RF AMPLIFIER CIRCUITS

A typical rf amplifier circuit designed for the AM band is shown in Fig. 5-3. Although it does provide some gain for the incoming signal, its main function is to provide image

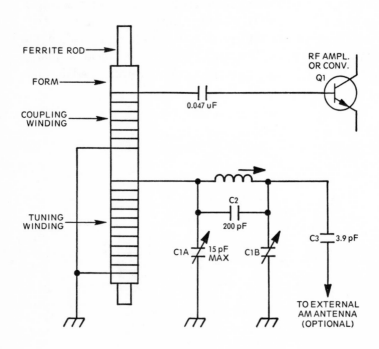

Fig. 5-2. Loopstick antenna circuitry.

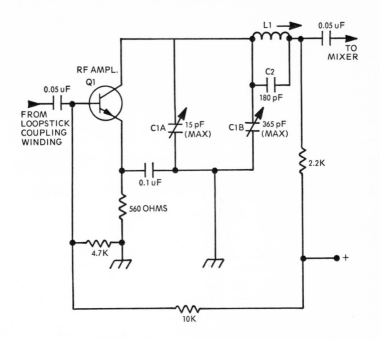

Fig. 5-3. Circuit of an AM rf amplifier.

rejection and increased selectivity. The theory here is that image signals will have less of a chance if they never even reach the mixer.

From a dc and bias point of view, the rf amplifier is much like any other common-emitter amplifier stage, no matter what the frequency range. In some receivers, however, one other type of bias is applied to the base. This dc level is derived from an automatic gain control (agc) circuit.

In most stereo receivers, the tuning for the base of the rf amplifier is performed by the circuitry associated with the loopstick antenna. Output tuning, however, is accomplished by the more familiar LC resonant tank circuit. In the circuit in Fig. 5-3, C1 (A and B) tunes L1 to the frequency of the desired station. An additional capacitor is sometimes used to shunt L1 to form a fixed-tuned parallel resonant circuit. This is used to block images on the high end of the band.

If this were a car radio design, where L1 is the variable tuning element, this circuit would offer superior image rejection over the entire band. For most home receiver ap-

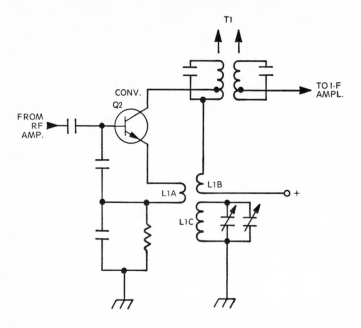

Fig. 5-4. Local oscillator circuit typical of those found in AM receivers.

plications, however, it is sufficient to offer additional attenuation only at the high end of the range because the tuned circuits become a bit less effective in that part of the band.

CONVERTER STAGES

A typical AM radio converter stage is shown in Fig. 5-4. This is only one of several dozen "common" designs. This particular circuit uses a three-winding local oscillator coil and is a hybrid between the tuned-collector and tuned-base type oscillators. In other converter designs, you may see either Hartley or Colpitts-Clapp oscillator circuits. In all cases, however, the collector of the converter transistor is connected first to the tap on the input i-f transformer primary winding. The impedance of the bottom half of this winding at local oscillator frequencies is low enough to be ignored. The local oscillator is only mildly aware of this inductance. However, it is a different story with the i-f signals produced by the mixing action. At this frequency, the impedance of the tuned i-f transformer is much more significant. The parallel-tuned i-f transformer selects the i-f signal while rejecting the two original signals and any other mixer products that happen to be present.

The circuit in Fig. 5-5 has become increasingly popular. This is called variously the "push-pull" or "differential amplifier" converter circuit. The i-f output signal is selected from the two collectors. One transistor serves to amplify the signal picked off the air by the loopstick antenna, while the other serves as the local oscillator.

Notice that the emitter resistor in this circuit is a bit larger than might normally be expected. Since the emitters appear at first glance to be in parallel, you might expect the common emitter resistor to be approximately one half the normal value. Instead, it is better than twice the normal value. This is because in the differential-amplifier type converter, the transistors must draw currents from a common constant-current source. This causes the two signals, local oscillator and rf, to divide the emitter current in a decidedly nonlinear

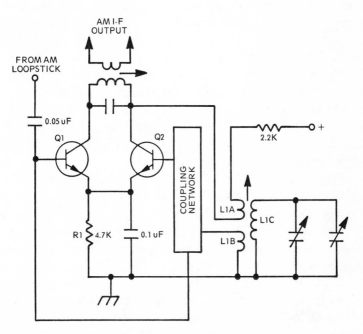

Fig. 5-5. Simplified schematic of the push-pull type local oscillator-mixer.

manner. This sets up the essential condition for heterodyning the two signals to form an i-f signal. If the circuit were linear, no mixing would take place and the resulting waveform would be one signal superimposed over the other.

I-F AMPLIFIERS

The i-f amplifier probably the single most important stage in a superheterodyne receiver, whether designed for AM, FM, or any other frequency band or mode of modulation. The i-f amplifier is primarily responsible for determining the reception specifications of the entire receiver. If this stage lacks gain, the receiver will be insensitive. If it also lacks selectivity, the overall receiver performance will be rather broad. These problems become really serious at night when long-distance atmospheric skip brings in signals from hundreds of stations over distances of up to a thousand miles or more. If the receiver has high sensitivity but lacks selectivity, the AM band at night will be alive with a large number of stations on the same and adjacent channels beating together to produce a regular cacophony of whistles, squeals, and buzzes. In this era, where we have a large number of AM broadcasters and sensitive radio receivers, we need that selectivity if we plan to use the receiver at all after nightfall.

The circuit in Fig. 5-6 shows a typical i-f amplifier using a bipolar transistor. The use of fixed-tuned transformers and custom-tailored bypassing allows the designer to maximize gain over the narrow range of frequencies passed by the transformers. This allows us to have both good selectivity and a high degree of signal amplification.

Modern receivers may use the integrated circuit (IC) approach in Fig. 5-7 to accomplish the i-f function. Special purpose ICs are available from almost every major semiconductor manufacturer. These units often perform far better than bipolar units at a comparable price level. The designer can approach the special purpose IC as merely a building block offering gain (sometimes a whole lot of gain). Input filters, such as tank circuits or ceramic crystals, can be used to select the narrow range of frequencies needed by the i-f amplifier. In the example in Fig. 5-7, ceramic crystal

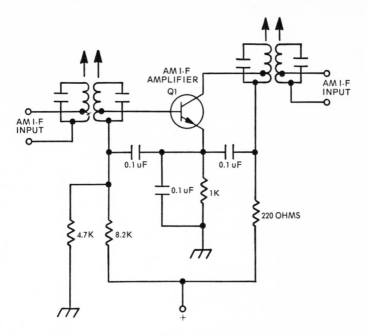

Fig. 5-6. Transistor AM i-f amplifier circuit.

filters, available in a variety of configurations and specifications, are used to provide i-f selectivity. The output of the IC is fed directly to the detector. As you can imagine, this allows a reduction in size, while offering the same or better performance.

In some receivers a separate crystal filter circuit is placed between the converter and the i-f amplifier gain package. In Fig. 5-8, we see a typical crystal filter circuit. Crystals Y1 and Y2 are electrically spaced a few kilohertz apart so that only a narrow range of input frequencies can pass. The precise frequency displacement depends upon the characteristics of the crystals and on the passband desired. In any event, the crystal filter passband is narrower than the passband of the i-f transformer at the circuit's input. In some cases, the crystals used in this filter are discrete components (as are the capacitors) and they are individually laid out on a printed-circuit board. In other cases, the entire circuit, except for the input transformer and the output terminating resistor, is contained in a small metal can about the size of one normal

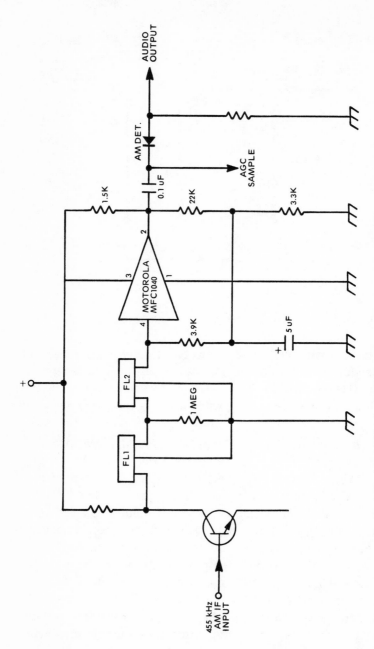

Fig. 5-7. Modern AM i-f amplifier circuit using crystal filters and an IC.

"miniature" i-f transformer mounted on its side. The value of the termination resistor is chosen to offer a reasonably decent match between the input of the amplifier transistor and the output of the filter.

AM DETECTORS

For AM detection, we need only a simple one-diode envelope demodulator circuit in order to recover the audio program modulating the carrier signal. Single-sideband reception calls for a product detector and for FM we need a frequency- or phase-sensitive detector. Therefore, the AM detector is the simplest of the detectors normally found in common electronic equipment.

In the typical AM detector (Fig. 5-9), a diode is connected to an impedance-matching tap on the secondary winding of the output i-f transformer. It is the function of the diode to provide the nonlinear environment needed to demodulate the AM signal. In this example, there are two series-connected filters following the AM detector. One is a pi-type RC filter designed to remove all rf and i-f components and to roll off the audio frequency response above 5 kHz. The impedance of the 470 pF capacitors decreases as the signal frequency goes higher. The

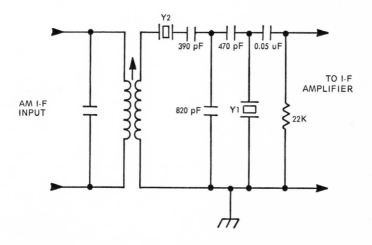

Fig. 5-8. Typical two-element AM crystal passband filter circuit.

impedance remains high, however, to audio frequencies within the desired passband. The RC filter effectively eliminates the higher audio frequencies and the rf/i-f components so that they cannot cause problems in the reproduced sound.

The parallel resonant trap, consisting of L1 and C3, is in series with the signal path. This trap is a 10 kHz suppressor. Its job is to suppress the heterodyne beats caused by the fact that AM band channel assignments in the U.S. are 10 kHz apart. If a signal from an adjacent-channel station gets through the i-f stages, it will beat against the desired signal to produce a difference frequency of 10 kHz. With tight i-f selectivity, a low-pass RC rolloff filter, and a 10 kHz parallel resonant trap, there is only a minimal chance of adjacent-channel heterodyne interference in the modern receiver. The price of good AM selectivity, though, is fidelity.

AUTOMATIC GAIN CONTROL (AGC) CIRCUITS

In FM receivers, we do not often find an automatic gain control (agc) system because most FM listeners choose

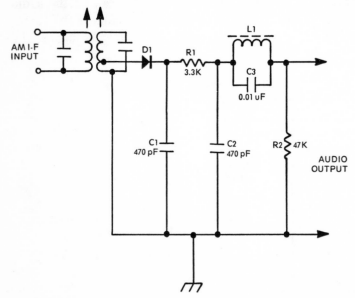

Fig. 5-9. Schematic of an AM envelope detector and 10 kHz whistle filtering system.

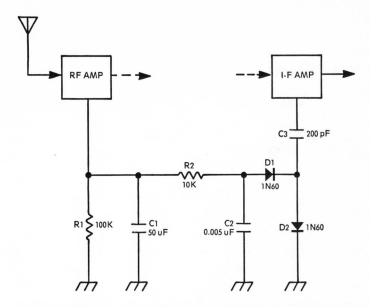

Fig. 5-10.Voltage-doubler AM automatic gain control (agc) circuit.

signals that are strong enough to drive the receiver into hard amplitude limiting. Indeed, it is this feature that, in part, accounts for the reputation for noise-free reception enjoyed by the FM medium. In AM receiver, on the other hand, there is a real need for agc. Without it, the listener would have to be continuously riding the gain control as he tuned across the band.

A typical agc circuit is shown in Fig. 5-10. In this arrangement, a capacitor-coupled sample of the AM i-f signal is fed to a diode voltage-doubling rectifier circuit. This creates a negative voltage across resistor R1. The resulting dc voltage, which is proportional to the signal level, is used to either raise or lower the gain of the rf amplifier according to signal strength.

Defects in this circuit can cause either heterodynes or a type of audio distortion unique enough to be dubbed "agc distortion." In that case, weak stations appear normal because they are all but unaffected by the agc system. Strong stations, however, drive the agc into conduction. This usually drives the rf amplifier transistor into either cutoff or hard saturation.

CHAPTER 6

Control Circuits and Antennas

To the new user, the fellow who was previously listening only to an AM table model radio, the variety of controls on the front of the modern FM stereo receiver may prove to be somewhat confusing. This effect may tend to increase in the future as manufacturers add more and more features which must be controlled. It should be helpful to take a look at some of these control functions, as well as the signal flow through a typical FM stereo receiver and some of the miscellaneous devices that contribute toward better system performance under marginal conditions.

PANEL CONTROLS

1.) **Volume control:** This one should present no problem. It performs the function the name suggests, and it is found on virtually all radio and TV receivers. It sets the absolute level of the sound produced by the receiver.

2.) **Loudness:** On most receivers, the loudness control is actually a switch. It modifies the tonal characteristics of the sound produced by the receiver at low settings of the volume control. This is provided because the linearity and sensitivity of the human ear changes at low sound levels. This phenomenon, by the way, is why many hi-fi salesmen turn the volume up to a high level when demonstrating a unit. A receiver is best auditioned at both low and high volume levels with and without the loudness control turned on.

3.) **Bass:** The bass tone control modifies the receiver's response to lower-frequency sounds. A typical control will vary the response between −15 and +15 dB, relative to 1000 Hz.

4.) **Treble:** This control is essentially the same as the bass control in overall function, except that it is effective in the higher-frequency octaves.

5.) **Mute:** The hiss level between stations on the FM band is extremely high. It is high enough and of such a character that it is extremely annoying. The mute control silences the output of the receiver unless there is a signal in the passband. Most of these controls work on the same principle as the "squelch" in communications receivers. A noise amplifier picks up the hiss from the FM i-f amplifier and uses it to generate a dc voltage that cuts off one of the stages between the FM detector and the stereo decoder. When a station is tuned in, the signal overcomes the noise (the receiver is said to "quiet") and thereby removes the dc cutoff bias from the controlled stage. When this occurs, the stage can function as a normal amplifier or emitter follower as the case may be. A "stereo only" mute kills the output of the receiver unless a pilot carrier is present in the FM multiplex decoder section.

6.) **Speaker selector:** There are usually two complete sets of speaker connections on the rear apron of the receiver, labeled something like "Remote: Right and Left" and "Main: Right and Left". The user might want to connect a pair of living-room speakers to the main connections and a pair of den or family-room speakers to the remote connections. The panel switch offers him the option of playing the main, the remote, or both systems.

7.) **Mono∕Stereo:** Some receivers have a provision for turning off the multiplex decoder. Although this may seem a bit like defeating the purpose of having an FM stereo receiver, it does have a purpose. The FCC rules allow a 10 percent maximum stereo modulation of the FM carrier; therefore, it is weaker than the monaural signal. At a distance, the L — R signal may prove to be too weak to allow proper decoding of the left- and right-channel components of the composite audio program material. As a result, the reception of stereo programing will be weak and noisy. The same signal may, however, produce a listenable monaural signal.

8.) **Input selector:** This control allows the front panel selection of the desired program source. It switches the input to the control amplifier between the FM, AM, phonograph, tape, and any auxiliary inputs.

SIGNAL FLOW

The recovered audio from the output of the demodulator is fed to a switch, which is either a bandswitch or part of the input or source selector, depending upon the specific equipment. From the switch, the signal is fed to the multiplex decoder and any audio buffers that might be in use. From this point on, the signal is stereo, so there must be separate amplifiers and a distribution path for each channel. The control amplifiers usually follow the buffer stages (Fig. 6-1).

In some receivers, there may be two mute gates. One is the normal mute gate, which detects the presence or absence of interstation hiss and noise. This circuit kills the output of the receiver in the same manner as the squelch circuitry in communications equipment. The only real difference is that the mute is controlled by a single switch and the mute level is set by an internal adjustment. On communications equipment, it is common to give the user access to the squelch level control. The stereo mute gate kills the output of the receiver unless the gate is opened by the 19 kHz pilot signal separated and amplified in the multiplex decoder.

SIGNAL LEVELS

It is impossible to establish absolute signal levels at the same points in all receiver, due to the wide variety of designs on the market. In general, though, we can make some statements about levels. For example, on the phono input, you can usually expect to find a signal with an rms level of less than 10 mV. This will drive the amplifier to full output. On the other inputs, expect to find that a couple hundred millivolts or more is required to accomplish the same job. One of the most common IC stereo decoders needs at least 250 mV of composite signal with 25 mV of pilot signal. In receivers using this chip, the specifications probably will adhere closely to these levels.

FM sensitivity is often specified as the number of microvolts of signal needed to produce a certain amount of

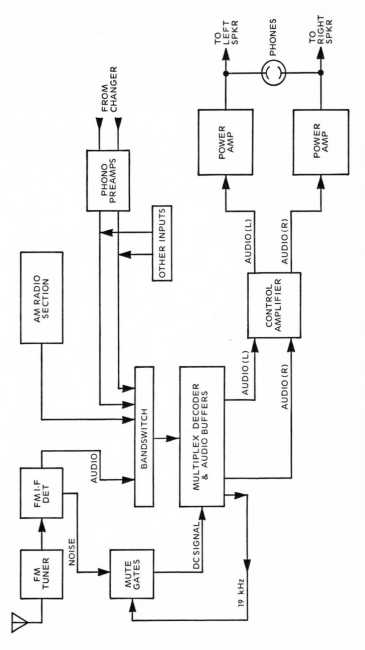

Fig. 6-1. Block diagram showing signal flow in a complex FM receiver.

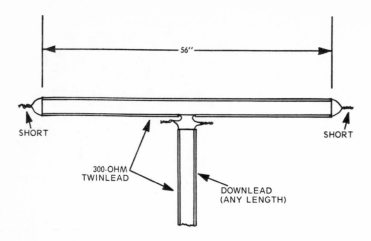

Fig. 6-2.Construction details for an FM folded dipole antenna.

"quieting" at a specific dial frequency. This is often called the signal-to-noise ratio. On the schematics of many FM receivers, there are notations designating signal levels at various points. These are usually the input and output terminals to the various sections of the receiver. Look in the small print somewhere on the schematic to determine exactly what conditions will produce these figures. In one typical case, you might find that the output of the control amplifier will show 800 mV rms. In the FM mode, the specification is realized when the receiver is connected through a 300-ohm transformer with a 1:1 voltage ratio at a 50-ohm generator that is supplying 100 uV of signal modulated by a 400 Hz signal to 22.5 kHz deviation at a dial frequency of 108 MHz. If you can set up these conditions on the bench, you will have a valuable aid to troubleshooting weak or marginal performing receivers.

ANTENNAS FOR HOME USE

The sketch in Fig. 6-2 shows a typical FM folded dipole antenna that is used with most FM stereo receivers. It can be made from 300-ohm twinlead, a type of TV antenna feedline. Cut one piece of twinlead to a length of 56 inches. Find the center of the lead and cut one of the two conductors; the other must remain intact. Strip the insulation from both ends of the cut wire. Connect the end of a second piece of twinlead (one lead each) to the two exposed wires at the center cut. This lead is the actual transmission line to the receiver and may be of any convenient length. Strip the insulation from the extreme ends of the 56-inch piece of twinlead and short the two exposed wires together.

This style antenna is usually supplied with most stereo FM receivers. It is also available ready-made from most electronic parts wholesalers and audio shops. The chief use for the homemade type, considering the low cost of the ready-made folded dipoles, is to make impromptu repairs while on a service call on console-type sets with ac line cord antenna troubles.

In a really bad area, it might be necessary to connect the receiver to one of the wideband FM yagi antennas available for rooftop mounting. Another alternative is to use a two- (or more) set coupler so that the FM receiver can be operated from the same antenna that serves the color TV set. Care must be exercised at this point, however, since many TV master antenna systems will not pass FM band signals. Many systems actually include traps in the antenna amplifier to suppress interference caused by strong FM signal intermodulation of the signals desired for TV reception. Such a TV antenna system severely attenuates FM signals. In still other cases, it might be necessary to use a variable attenuator in the FM receiver feedline because the TV amplifier boosts the FM signal too much and causes severe intermodulation problems. In a system affected by this problem, the stronger stations probably will appear at several points on the FM dial!

AM BAND ANTENNAS

In some areas of the country, especially those that are remote from the large metropolitan areas, it may be necessary to improve AM reception by the use of an external antenna. On some receivers, this is a simple process because some manufacturers provide a terminal for the connection of such an antenna. In other cases, it might be necessary for the user or a service technician to provide such a feature.

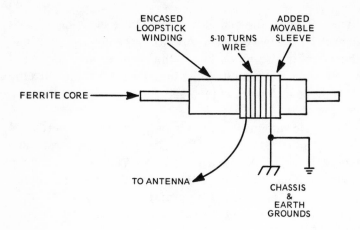

Fig. 6-3. This is one method of connecting an external AM antenna.

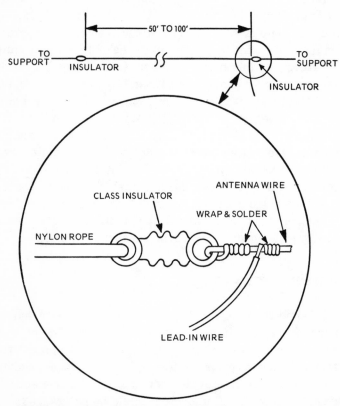

Fig. 6-4. Construction details for an external AM long-wire antenna.

One simple way to make an external AM antenna work on most receivers is illustrated in Fig. 6-3. Fashion an insulated sleeve that fits over the loopstick. It should fit just tight enough so that you can move it by pushing, but snug enough so it won't move on it's own. Wrap five or ten turns of enameled or plastic insulated hookup wire (anything in the No. 22 to No. 26 range should do fine) around the movable sleeve and secure with coil or Q dope cement. Connect one end of this homebrew coil to either earth or chassis ground and the other end to the external antenna.

A suitable antenna for use with the connection in Fig. 6-3 is illustrated in Fig. 6-4. The length should be over 50 feet, but try to keep it under 150 feet, since antennas of large size are difficult to manage. Many radio parts houses sell these antennas already made up for a price under $10 list.

Should overload on local stations prove to be a problem, it may be necessary to disconnect the antenna when tuned to such stations. In some cases, connecting a 150 pF variable capacitor in series with the antenna at the back of the receiver will give enough attenuation on local signals. Then, the capacitor can be readjusted for more distant reception. Receivers in a metropolitan area will probably be subjected to severe overload all across the AM band if an external antenna is used.

Power Supply Circuits

It can be said that the power supply provides the "lifeblood" for an FM stereo receiver. In large measure, the functioning of the power supply determines at least one of the real specifications of the entire receiver. Take the matter of audio power output, a figure that users and salesmen like to boast about. It is common practice to rate stereo receivers according to a "watts per channel" system. A certain stereo receiver might be able to deliver, say, 50W of rms audio power from each channel. Is it, then, proper to say that a two-channel receiver will produce a total of 100W (2 x 50) and that a quadraphonic receiver will produce 200W (4 x 50)? Unfortunately, this is not necessarily the case. The limiting factor regarding such figures is the power supply output capability. If that power pack can deliver only 90W of power (dc) without exploding or doing something else equally catastrophic, we should claim the receiver's power level to be only 45 watts per channel **with both channels driven to maximum**. A really well designed amplifier will actually be able to deliver slightly in excess of its rated power. A receiver offered as a 50W per channel unit should have a power supply capability of over 100W total.

POWER SUPPLY REQUIREMENTS

You will see many different types of power supplies in stereo receivers. The typical power supply, however, consists of a transformer to modify the 110V 60 Hz ac obtained from the power mains to a level of voltage and current consistent with the needs of the receiver's solid-state circuitry.

Following the secondary of the transformer is a solid-state rectifier system used to convert the ac at the input to a form of pulsating dc. These pulsations are smoothed out to a pure dc by the filtering system following the rectifiers. In some applications, there might also be a voltage regulator system for the lower-power-level stages. For the higher-power stages, voltage regulation requirements must be met by an abundant current capability. The result is a lower than absolutely necessary internal impedance requirement for the power supply.

The absolute dc levels used in stereo receiver power supplies tend to vary from brand to brand and from model to model. In general, however, the lower-power units operate with a voltage level under 25V dc and a current capability of less than 1A. Higher power receivers contain power supplies capable of producing over 100V at a couple of amperes. Most receivers, however, operate with power levels in between these two extremes, somewhere around 40-60V.

Most receiver power supplies fall into either the single-pole or bipolar category. In the single-pole type, either the positive or negative terminal is grounded to either the chassis or to a floating counterpoise ground that also serves as signal common. In the bipolar power supply, there are actually two intermingled power supply circuits. One produces a dc voltage that is negative with respect to the chassis or counterpoise ground, while the other voltage is positive with respect to ground. This type of supply is almost mandatory for circuits using an operational amplifier IC in the preamplifier stages and for certain types of capacitorless and transformerless power amplifier stages.

POWER SUPPLY CIRCUITS

A typical power supply for a receiver in the low-power class appears in Fig. 7-1. This is a simple two-diode full-wave rectifier circuit. It produces a much smoother form of pulsating dc than does the simpler half-wave circuit used in low-cost table model FM radios. The ripple frequency of a half-wave circuit is equal to the line frequency (60 Hz), while

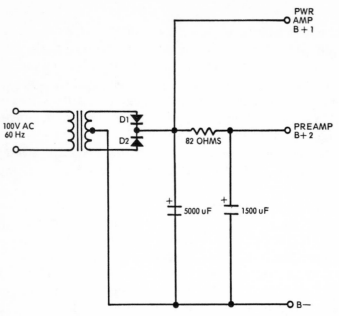

Fig. 7-1. Traditional full-wave rectifier power supply circuit.

the ripple frequency of the full-wave circuit is equal to twice the line frequency (120 Hz). A higher ripple frequency means that less is required of the filtering network.

Consider what happens to a sine wave applied to the primary of the power transformer. It has a voltage level between the limits of 105 and 125 and a current level as required by the design of the transformer and the requirements of the receiver. The voltage and current levels are modified to those actually needed by the receiver circuits in the secondary circuit of the transformer.

On one half of the input sine wave, diode D1 conducts and produces a half-wave pulse across the filter capacitor. During this interval, the voltage presented to diode D2 is equal in amplitude but of opposite polarity to that seen by the other diode, D1. Consequently, D2 is reverse biased and cut off. When the input sine wave polarity reverses, the opposite situation occurs; D2 becomes forward biased and D1 is cut off. This causes a second half-wave pulse to appear across the filter capacitor immediately following the pulse produced by D1. In a half-wave system, there is a time interval between pulsations equal to one half the period (reciprocal of frequency, of 1/60) of the input sine wave. Pulsations at this frequency are more difficult to filter out than are the pulsations of a full-wave circuit.

Notice the high value of the input filter capacitor. In solid-state receivers, the value of this component ranges from 2,000 to 12,000 uF, depending on the design. Since dc voltage levels across these capacitors can approach 100V in some receivers, extreme care must be exercised by service personnel. These sets are not portable transistor radios with the comparatively docile low-voltage levels provided by batteries. In those sets, we tend to get a little careless because we can touch anything in sight without any chance of injury. In some modern solid-state stereo receivers, it is mandatory that we exercise the same degree of caution that we observe in older tube-type receivers.

In the power supply circuit in Fig. 7-1, a voltage-dropping resistor and additional filtering are used to supply the power to some of the lower-level stages preceding the power am-

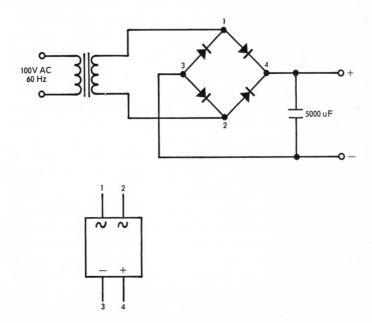

Fig. 7-2. Four-rectifier bridge power supply circuit.

plifier. The second filter capacitor serves to decouple these stages from other receiver circuits and further suppress the ripple component.

The power supply circuit shown in Fig. 7-2 is a full-wave bridge circuit. Notice the absence of a center tap on the power transformer secondary winding. This circuit has the ability to produce a dc voltage output that is roughly twice the output of the same transformer feeding a simple full-wave rectifier. This is because the bridge makes use of the entire secondary winding during each alternation, rather than just one half the total winding. Of course, since a transformer cannot generate power, only convert what it receives, the bridge circuit also delivers one half the current capability of the two-diode full-wave system.

The bridge is probably the most popular power supply configuration in stereo receivers. It uses one pair of the four diodes on each half of the input sine wave. One pair of diodes is active on one alternation, while the other pair is active on the opposite alternation.

Diodes in the bridge connection are found in a variety of styles. In many receivers, the bridge is constructed of four separate and independent diode rectifiers. In others, you will find one of many available bridge-connected four-diode packages. One common type of diode package is illustrated in the inset at Fig. 7-2. This package is a thin, square sandwich with the diodes mounted inside. The terminals are marked with the "sine" symbol for the ac connections and with "+" and "−" for the dc connections.

There are numerous other bridge rectifier stacks used in stereo receivers. In many instances, it takes some imagination to identify certain imported rectifier stacks. In these cases, only the fact that the power transformer leads are attached to one set of terminals and the filter capacitor to the other set indicates that the component is a rectifier. In a few cases, there are only three leads attached to the bridge stack. In this situation, you will also notice that the stack is bolted to the common point or chassis. The bolt is the terminal for the negative side of the dc output. It remains grounded as long as the stack is bolted into place.

The circuits requiring a bipolar or dual-voltage power supply can be served by the configuration illustrated in Fig. 7-3 if the power-level demand is low enough. In this arrangement, the dual-voltage characteristic is obtained from two series-connected zener diodes shunted across a normal monopolar power supply. The junction of the two diodes serves as the common ground and is connected to either the chassis or a counterpoise system. In most applications, there are two filter capacitors across each zener diode. One is the normal high-value electrolytic used for low-frequency decoupling and ripple elimination, while the other is a low-value capacitor of another type for suppression of higher-frequency parasitic oscillations.

The circuit in Fig. 7-3 is only useful where bipolar operation is required only in the lower level stages, for example, where an operational amplifier used as the active element in one of the preamplifiers or the control amplifier. In receivers where the power amplifier is operated from a dual-

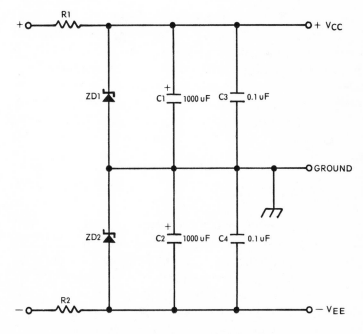

Fig. 7-3. Zener-derived dual-polarity power supply circuit.

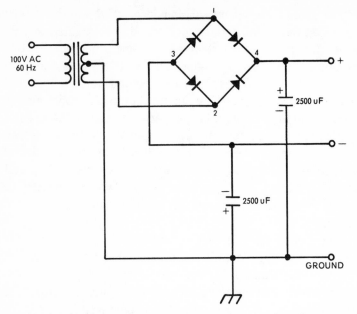

Fig. 7-4. Dual-polarity power supply circuit using a transformer and bridge rectifier.

voltage supply, a circuit such as that shown in Fig. 7-4 is used. In this configuration, there is a center-tapped transformer secondary and a standard bridge rectifier stack. The bridge functions as two independent half-wave circuits rather than as a full-wave bridge. The center tap on the power transformer secondary establishes the common ground for the system. Each half of the dual-voltage supply has its own filtering network. Again, as in previous examples, the values of the filter capacitors tend to fall in the 2,000 to 12,000 uF range.

VOLTAGE REGULATION

The power amplifier must depend upon reserve current capacity to furnish dc regulation. In some of the lower-level stages, however, a tighter degree of regulation can be afforded by relatively low-cost electronic regulator circuits. This tighter regulation is especially desirable on the power line to the FM tuner because variations in this dc level can deteriorate the stability of the entire FM function.

The power supply circuit in Fig. 7-5 shows two basic voltage regulator circuits used in present stereo equipment. One type is the simple zener diode-resistor regulator consisting of ZD1 and the 82-ohm current-limiting resistor. The zener diode has the ability to maintain a relatively constant voltage drop in the reverse direction in the face of rather severe excursions of the input level.

The other dc regulator employs the current amplification factor of a power transistor to increase the regulating ability of the supply. This is the type of circuit generally used to supply power to the FM tuner, while the circuit featuring only a zener diode can be used to feed those stages where the degree of regulation is not so critical. The base of the transistor is held at a constant voltage by zener diode ZD2. If an increase or decrease in the output load forces the emitter voltage up or down, the resulting change in the base-to-emitter bias causes the collector current of the transistor to readjust, which maintains the output voltage at a constant level. A post-

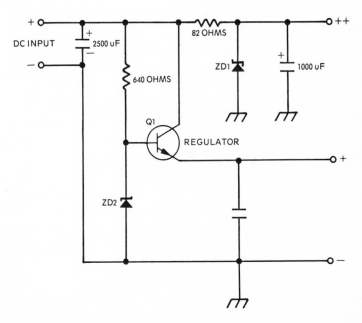

Fig. 7-5. Schematic of a voltage regulator using a transistor and zener diode.

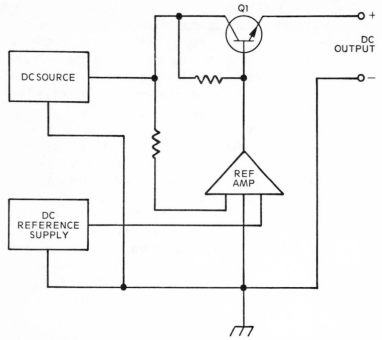

Fig. 7-6. Schematic of an electronically controlled "reference amplifier" regulated power supply.

regulator filter capacitor minimizes the abruptness of these changes and aids in the further attenuation of the ripple component.

The regulated supply in Fig. 7-6 is one of the more modern circuits found in only a few of the most recent FM receivers. The series-pass regulator transistor is controlled by the output of a differential amplifier, which is often an IC op-amp.

One feature of a differential amplifier is that it responds only to the difference between two input signals. When both inputs are at the same level, or both are zero, the differential amplifier will allow no output. In that case, the total base bias current for the series-pass regulator is supplied by the collector-to-base resistance.

One of the inputs to the differential amplifier is a zener-regulated dc reference voltage. This voltage is usually supplied through a chain of voltage-dropping resistors and zener diode regulators to assure maximum stability. This allows a greater degree of regulation than would be possible if the same multiple zener method had been used as the primary method of regulation because the output of this zener feeds a fixed load.

The other input to the differential reference amplifier is a sample of the dc power supply output voltage. Depending upon whether the output load current increases or decreases, the differential amplifier will either add to or subtract from the base current of the series-pass transistor. This, in turn, causes the transistor to turn on harder or to go toward turnoff. The result is a relatively constant dc voltage level. If the manufacturer is willing to spend the money, this technique can even be used to regulate the level of the voltage fed to the power amplifier.

Stereo Measurements

Occasionally, it is necessary for the service technician to make certain measurements on FM stereo receivers. Specific checks and tests are used to uncover defects, to verify the condition of a receiver prior to delivery to a customer, or to check performance on customer demand. Therefore, it is incumbent on the service technician to be able to make these measurements in an accurate and proper manner.

POWER OUTPUT

The output power is one of the most frequently misquoted, cussed, discussed, and abused high-fidelity power amplifier ratings. Contributing to the confusion is the fact that there are several methods for determining amplifier output power and they all yield vastly different numerical results. The particular method used seems to be related to exactly what the measurer is trying to prove. The confusion and chaos is so great that some technicians are quoting "so many watts IYL." The abbreviation "IYL" means "If You're Lucky." There are several qualifications that must be made in any power measurement before the result can be rationalized.

The test setup for a wide range of amplifier measurements is shown in Fig. 8-1. It is necessary to have a source of audio frequency sine waves with a distortion figure at least one order of magnitude better than that offered by the amplifier being tested. If the amplifier is rated at 1 percent total harmonic distortion (THD), the audio oscillator must have less than 0.1 percent THD. Because of the extremely low levels of THD offered by modern stereo equipment, the technician must equip himself with only the best audio sources available.

Whatever audio source is chosen, there must be a means for reducing the output to a level that will properly drive most amplifiers. In most cases, the audio source includes a calibrated attenuator. Even though the attenuator is billed as being accurately calibrated, it is desirable to include an ac vtvm in the amplifier input circuitry to measure the absolute rms value of the input sine wave.

The output of the amplifier is loaded with a "dummy" speaker, an 8-ohm noninductive resistor of an appropriate power level. Considering the claims of many modern receivers, it seems like a 50W dummy load is the minimum

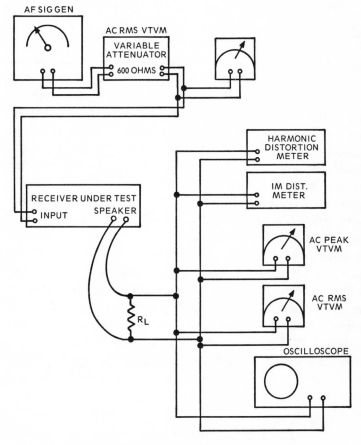

Fig. 8-1. Bench setup for general measurements.

acceptable, while a 100W load is at least highly desirable. Mounting a 50 or 100W resistor of appropriate design to a heatsink or in an oil bath effectively increases its apparent power rating.

Few noninductive resistors are wirewound. The cross-sectional diameter of the wire element in the resistor is one primary factor in determining a resistor's suitability as a dummy load. The current developed by the amplifier must not exceed the current capability of the wire. The resistors encased in a flanged metal housing are preferable because of the power rating increase realized by mounting it to a finned metal power transistor heatsink and by forced air cooling with one of the newer types of "whisper fans."

The audio test equipment is connected in parallel across the dummy load (Fig. 8-1). Of particular use are: a total harmonic distortion analyzer, an intermodulation distortion analyzer, an rms ac vtvm, a peak-reading ac vtvm, and an oscilloscope. It is more efficient to provide a bank of switches so that the instruments can be connected into the circuit at will. Another switch can also be provided to replace the dummy load with a real speaker system. This arrangement can result in an extremely efficient test bench. If appropriate, another switch can be provided to transfer the entire collection of test instruments to the opposite channel. Do not, however, include the dummy load in the transfer operation. Each channel should have its own load, regardless of where the instruments are connected.

To make the power test, turn the volume control on the amplifier to maximum and apply a signal of an appropriate voltage level from the audio generator. Turn the signal up until the rated THD figure is reached. If the specifications claim that the power rating was measured at 1000 Hz and 1 percent THD, turn the input voltage up until the 1 percent THD figure is reached. At this point, measure the rms ac voltage across the dummy load. The power output of the amplifier is equal to:

$$\text{Power (watts)} = \frac{(\text{rms voltage})^2}{(\text{resistance of dummy load})}$$

The dummy resistance in most cases is 8-10 ohms.

Peak power, a rating preferred by some manufacturers because it tends to make their equipment look more powerful in the eyes of the less sophisticated buyer, is obtained from the equation:

$$P = \frac{(E \cos \pi FT)^2}{R_L}$$

Where P is the peak audio power output in watts, E is the peak voltage of the output sine wave as measured on the peak ac vtvm or oscilloscope, F is the test frequency, and T is the time in seconds for each cycle.

If any particular power level is stated, it must be qualified by the level of harmonic distortion and by the number of channels operating from a common power supply. It is common to find amplifiers rated with only one of the two channels driven to full output at a time. If the specification sheet calls an amplifier capable of producing 50W per channel, but the power supply can only deliver a total of 80W, it would be more honest to label the amplifier as a 40W per channel unit, not 50W. Unfortunately, pressures of the marketplace often dictate that the single-channel power be advertised to keep from having a "less scrupulous" competitor scoop the sales potential due to his allegedly more powerful product.

If the total harmonic distortion level is not stated in the advertising, it may be that the specified output power cannot be achieved at normal THD levels. An amplifier that can produce only 20W at 1 percent THD may well be able to produce upwards of 50W with a THD of 5 percent instead of 1 percent. It is less than honest to use the 5 percent figure unless it is so stated. In this author's opinion, a THD of 5 percent, considering the state of the art, does not qualify an amplifier as "high fidelity." In fact, 5 percent THD is rather low "fidelity."

Another consideration, in a few cases, is the length of time an amplifier can sustain its so-called rated power level. If that time is for only a fraction of a second, the power rating is meaningless. Most quality high-fidelity receivers will operate up to five minutes at full power. Power level comparisons must be made at the same frequency, at the same THD figure, and with the same number of channels driven. You will notice that many modern amplifiers fall down when power is measured with all channels (two or four) driven to the maximum level simultaneously. This is because the power supply capability becomes significant under these conditions.

FREQUENCY RESPONSE

The test setup used for power output measurements can also be used to make a step-by-step check of amplifier frequency response. Start at 1 kHz with the amplifier producing either full rated output or 1W, or whatever power level the manufacturer intended, with a THD figure at or less than 1 percent. Use the rms signal voltage output under these specified conditions as the 0 db reference level. Check the output voltage level at least every octave (an octave is a two-to-one frequency spread; 1000 Hz is one octave above 500 Hz and one octave below 2000 Hz) and preferably more often. This must be done while maintaining a constant signal input voltage level. The resultant output voltages can be used as the data to construct a frequency response curve calibrated in either absolute voltage terms or in relative decibels related to the 1000 Hz figure.

If quantitative information is not required, the square-wave method offers a quick and convenient method of measuring frequency response. A square wave is fed to the input of the amplifier being tested. It must be noted that since the input waveform is not sinusoidal, the ac vtvm is of limited usefulness during this test. It is necessary to use an oscilloscope to determine proper input and output voltage levels.

One property of a perfect square wave is that it contains many harmonics and subharmonics of the fundamental frequency. The rule of thumb usually stated is that a square wave will indicate the frequency response of an amplifier over a range of from one tenth to ten times the fundamental frequency. A 1 kHz square wave, therefore, can be used to rough check the response from 100 Hz to 10 kHz. A 5 kHz square wave is good for response checks up to 50 kHz. It must be noted, however, that the use of too high a fundamental frequency may tend to give apparently erroneous results due to the amplifier's normal rolloff at higher than audio frequencies. An amplifier flat to 20 kHz may be downgraded because of a poor waveform in a square-wave test if the input frequency is over 2 kHz. This isn't critical, but it can cause problems.

A normal square wave is shown in Fig. 8-2A. In Fig. 8-2B, we see what happens to the waveshape if it passes through an amplifier or other circuit that attenuates the high-frequency components. You'll see this waveform if an amplifier has poor high-frequency response. If the waveform in Fig. 8-2C is present, we can assume that the low-frequency response of the amplifier is affected. Of course, if the wrong frequency is chosen for the fundamental, either of the two apparent defects might show up. For example, if the input frequency is 10 kHz or more, expect a high degree of high-frequency attenuation.

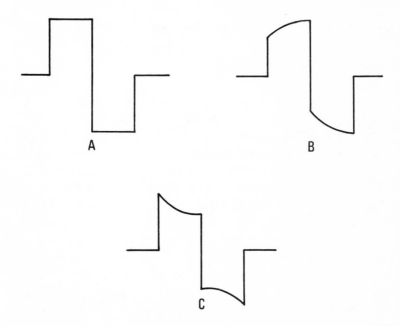

Fig. 8-2. Square waves indicate the frequency response of an amplifier. Waveform B shows a loss of high frequencies, and C indicates a low-frequency loss.

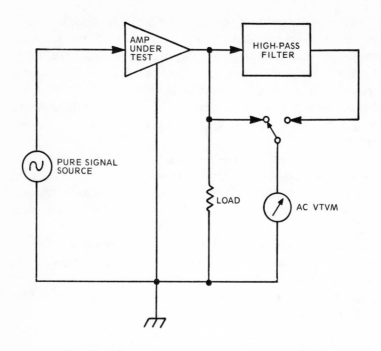

Fig. 8-3. Harmonic distortion analyzer test setup.

AUDIO INPUT SENSITIVITY

The bench test setup in Fig. 8-1 can also be used to check the input sensitivity of an amplifier. The sensitivity is the rms level input signal required to produce either the full-rated output or a certain specified output power. If the amplifier is rated to produce, say, 20W with a 250 mV input signal at 1000 Hz, we may properly say that the amplifier has a 0.25V sensitivity. All inputs should be compared at the same output power level, and the THD at this power level should be no more than 1 percent or the manufacturer's specified level if less than 1 percent.

Another valid test might be to rate an amplifier's sensitivity at an output power of 1W. The input sensitivity ratings of two different amplifiers can be compared by the proportion:

$$\frac{V1}{V2} = (\frac{P1}{V2})^{1/2}$$

TOTAL HARMONIC DISTORTION (THD)

The basic test setup in Fig. 8-3 is used to measure total harmonic distortion in an audio amplifier system. An rms ac vtvm is used to measure the output signal voltage of the amplifier under test. After this measurement is taken, the signal is fed through a high-pass filter that attenuates the fundamental while permitting the harmonics to pass. The voltmeter is then used to measure the rms value of the remaining signal. This is the harmonic content. If the test set is arranged so that the output of the amplifier drives the meter to exactly full scale, we can make the voltmeter read in percentage THD. We do this by adjusting the fundamental signal level so that the meter reads exactly full scale. When we switch in the high-pass filter, the reading represents only the harmonics. If the meter scale is calibrated in percentage, the THD is indicated directly.

Another means of making a quick check of the THD is outlined in Fig. 8-4. In this method, we use an oscilloscope and an audio generator. The output of the audio generator is fed to

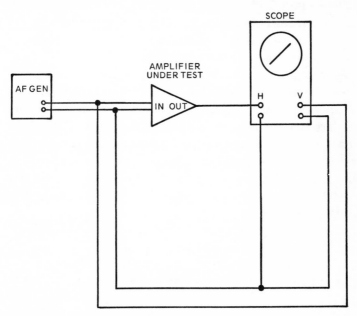

Fig. 8-4. Harmonic distortion test setup using an oscilloscope.

both the amplifier input and the vertical input of the oscilloscope. The output of the amplifier is fed to the horizontal input of the oscilloscope. The scope input controls are adjusted to bring the trace on the screen in a viewable manner. The result will be a relatively straight line that indicates both THD and relative phase shift. In the case of THD, it is the curvature of the line which is of interest. If the output of the amplifier is identical to the input (zero THD, an impossible but approachable situation) the trace line will be absolutely straight. The existence of harmonic distortion will cause the tips of the line to bend, one way or the other. The angle of the line relative to the baseline will give phase-shift information.

CHANNEL SEPARATION

In all stereo amplifiers and receivers, there is a certain amount of signal crossover from one channel to another; that is, a small amount of the signal in the right channel will appear in the left channel output and vice versa. In a real amplifier, the only way to eliminate this problem altogether is to completely isolate the circuitry of the two channels. This also means separate power supplies and separate cabinets. Of course, all of this effort would amount to nothing, since no perfect stereo program material is available.

The degree of separation, or lack of it, can be measured. Since it is indicated by relative rms signal voltage levels, we can express the relationship in decibels. The most common procedure is to connect the equipment as shown in Fig. 8-5. In the case of an amplifier, you would use an audio generator in place of the FM stereo multiplex generator and would feed the input to only one of the two channels. The two rms ac meters indicate the signal voltage output for each channel. The separation is equal to:

$$\text{Separation (dB)} = 20 \log \frac{E_{\text{channel 1}}}{E_{\text{channel 2}}}$$

where channel 1 is the channel to which the input signal is being delivered and channel two is the channel that is supposedly dead.

Some specialized test systems built by audio servicers for their own use include but a single meter. These systems have a level control so that the reading of the live channel can be adjusted to deflect the ac vtvm to exactly full scale. A switch is provided to transfer the meter to the quiet channel. The meter scale is calibrated in decibels so that the arithmetic is performed automatically in the process of the measurement.

In cases where you want to measure or adjust the separation offered by the stereo decoder section of an FM stereo receiver, it is necessary to use a multiplex generator. Some generators allow direct connection to the decoder input via a composite output cable or jack. This author prefers to use a system where the composite signal modulates the output of an FM signal generator. The generator output is connected to the receiver's antenna input circuit, which allows you to check the entire receiver at once. This is advisable because some FM i-f amplifier problems can cause a loss of separation, and in some receivers it is possible to over-attenuate the pilot signal in the RC network present in most FM stereo decoder input sections. The FM multiplex generator should be of the type that can produce a "right

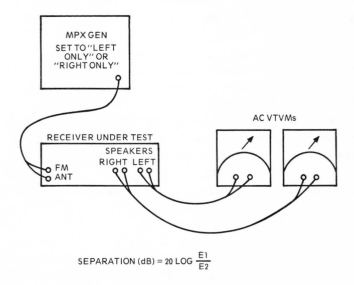

$$\text{SEPARATION (dB)} = 20 \text{ LOG } \frac{E1}{E2}$$

Fig. 8-5. Stereo channel-separation test setup.

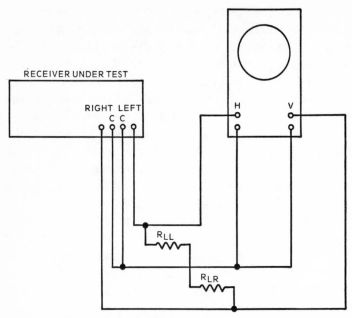

Fig. 8-6. "X-Y" scope separation test setup.

only" or "left only" composite modulated by a 1000 Hz audio tone.

The test setup for a relatively unknown method of measuring channel separation is shown in Fig. 8-6. With this method, it is also necessary to use either a real "X-Y" oscilloscope or a regular scope with the vertical and horizontal input sensitivities equalized. It is important that equal voltages applied to the two inputs cause equal deflection in both directions. In other words, if a 0.1V rms signal to the vertical amplifier causes a 1 centimeter deflection, the same signal applied to the horizontal input should also cause a 1 centimeter deflection in the opposite plane. If the signal is fed to both inputs simultaneously, the trace will be deflected in both planes an equal amount, resulting in a straight line positioned at a 45 degree angle. This test is less valid if the scope isn't calibrated.

The electrical requirements of the oscilloscope are not high. In fact, some technicians who use this system have found that almost any oscilloscope is suitable. Many shops still have an old audio response recurrent sweep oscilloscope stuffed

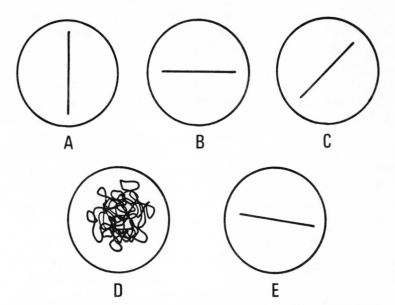

Fig. 8-7. Waveforms encountered during the test setup in Fig. 8-6. See text.

under a bench someplace. A considerable savings in troubleshooting time can be realized if the scope is connected across the speaker lines as shown at all times during troubleshooting.

The drawings in Fig. 8-7 show some of the traces you may find in using this measurement technique. The trace at Fig. 8-7A will appear if only the right channel is producing an output. If the left channel is the "live" channel, you will see the trace in Fig. 8-7B. In these two traces we see the result of perfect channel separation. If a monaural program source is fed to both channels with equal amplitude, we will see a trace similar to Fig. 8-7C. The 45 degree angle of the trace with respect to the baseline indicates that both scope inputs are being fed with the same amount of signal. In most cases of poor separation, the trace is something between one of the perfect traces and the monaural trace. The less than perfect stereo trace will appear at some angle other than 45 degrees.

The above traces will appear with either a sine-wave input to an audio amplifier or sine-wave modulation of an FM stereo signal generator. In the trace of Fig. 8-7D, we see the trace to be expected when tuned to a station with good stereo separation. This type of trace has been described as a "mess of very angry snakes" or a "plate of spaghetti that doesn't want to be eaten." It is continuously in motion as long as there is modulation present.

This scope is extremely useful in aligning certain IC stereo decoders. In those sets, the 38 kHz signal is of such a low amplitude that normal alignment by the peaking method is almost impossible. With this method of alignment, we need only adjust the 38 kHz transformer while the receiver is tuned to an FM signal generator output signal that is modulated on only one channel. Tune the slug of the 38 kHz coil until the trace lies as close to one axis as possible. Be sure, of course, that it is the proper axis. In most receivers employing the Motorola designed FM stereo decoder IC, it is possible to completely reverse the stereo channels by reversing the phasing of that 38 kHz coil! If you adjust the coil as described, the receiver should exhibit maximum separation.

If the oscilloscope is left connected to the bench speakers, you will have a means for determining whether a "no stereo lamp" complaint is the fault of the lamp circuit or the receiver's miltiplex circuitry. If the "mess of snakes" pattern is present when you tune to a stereo station but the lamp is not lit, it is evident that the trouble is in the lamp or its associated circuitry. On the other hand, if the trace is missing or if one of the monaural traces is present instead, look in either the FM receiver circuits or the multiplex decoder circuits.

FM RECEIVER MEASUREMENTS

One of the primary FM receiver specifications is sensitivity. This is a measure of the receiver's ability to respond to weak signals. Unfortunately, there are almost as many ways to measure and express the sensitivity figure as there are ways to measure output power. This gives the advertising people widely different figures to play with.

One of these methods involves measuring the number of microvolts of signal required to produce a given output at the FM detector. In this test, the signal generator, modulated by a 400 or 1000 Hz tone at either 22.5 or 75 kHz deviation is connected to the receiver antenna terminals through the suggested pads or matching networks. The output level of the signal generator is brought up until the specified output voltage (rms) has been developed. The number of microvolts needed to produce that level can then be read directly from the signal generator's attenuator dial. A typical specification might read: 1.8 uV produces 350 mV when modulated 100 percent (75 kHz) by a 1000 Hz sine wave.

Another popular method involves measuring the amount of signal required to produce a specified amount of "quieting." The first step is to read the full-volume output noise (no signal) on an ac vtvm. The signal level of the generator is then raised from zero until the output noise drops a specified amount (usually one tenth or 20 dB). The signal in this case must be unmodulated. This is called the signal-to-noise ratio. A typical specification might read 1.2 uV for 20 dB of quieting.

The capture ratio of a receiver can mean one of two things, depending upon whom you talk to. In one case, it refers to the well known capability of an FM receiver to reject weaker signals that are on the same channel with a strong signal (co-channel ratio). In another context, it refers to the tendency of strong signals on adjacent channels to capture the afc.

In the case of co-channel capture ratio, we use two FM signal generators set to the same output frequency. One is modulated 22.5 kHz by a 400 Hz sine wave, while the other is modulated a like amount by a 1000 Hz sine wave. The 400 Hz generator is set to some specified output level (100 uV is a common value). The output of the other generator is increased from zero until it overpowers and completely suppresses at 400 Hz tone. The ratio of the 1000 Hz generator's output at the capture point to the reference output of the 400 Hz generator expressed in decibels is the capture ratio.

In the other type of capture ratio measurement, we do exactly the same thing, only the two signal generators are separated in frequency by the width of one FM broadcast band channel, which is 400 kHz. The result of the second type of capture ratio test is a relative indication of the receiver's freedom from adjacent-channel interference. In this case, the higher the capture ratio the better the receiver.

Airplane flutter can be simulated by applying subsonic (5 Hz or so) amplitude modulation to an FM signal. If you record the percentage of 5 Hz amplitude modulation present when you first become aware of the flutter, you will have an indication of how free the receiver is of interference from multipath conditions caused by overflying aircraft.

The bandwidth necessary in an FM stereo receiver seems to be open to some question. Some authorities claim that 200 kHz (2 x 75 kHz deviation plus 2 x 25 kHz guard bands) is sufficient. Others claim that because of the modulation index allowed and the maximum modulating frequencies, there is a total of eight significant Bessel function sideband pairs present. This means a required bandwidth of over 300 kHz.

In any event, it is occasionally necessary to measure the bandwidth of an FM receiver. This is done by connecting a sweep generator capable of sweeping across at least 600 kHz of the FM band at a sweep rate of 60 Hz. The horizontal oscilloscope sweep is synchronized to the sweep rate of the generator (either by external sync from the generator or by internal connection to the ac line frequency), while the vertical channel is connected to the output of the circuit being checked. The result should be the familiar bell-shaped curve.

You can find the center of the passband by adding a 10.7 MHz marker. The marker should, in the interest of accuracy, be crystal controlled. Additional markers can be produced by using a 100 kHz marker crystal. This will show the relative response at the 10.5, 10.6, 10.8, and 10.9 MHz points. The receiver manufacturer usually specifies the useful bandwidth as the response measured between the half-power or half-voltage points, which is 6 dB down for voltage and 3 dB down for power. In most designs, the 6 dB voltage points are just over 200 kHz apart.

Test Equipment for Stereo Servicing

CHAPTER 9

No matter how you look at it, modern stereo equipment must be considered as high-quality electronics. Because of this fact, the technicians who service stereo and audio systems must, of necessity, be equipped with the best test equipment. It is useless to measure, say, total harmonic distortion with an analyzer whose internal THD is several orders of magnitude higher than the equipment being tested! Unfortunately, test equipment in this class tends to be a bit expensive. In the long run, however, it will more than pay for itself.

If you are a retail sales dealer, it is wise to perform a predelivery checkout of stereo units. This is reassuring to the purchaser of what has become a major appliance in the modern home. This is especially true if he is a novice hi-fi customer who has only the minimal knowledge necessary to risk a glance at a specification sheet. As another customer service, many dealers are offering, for a fee, a "diagnostic analysis," similar to the service provided by those "scientifically equipped automotive diagnostic centers." At least one shop of this author's acquaintance maintains a card file which contains a record of the warranty expiration date of each unit he sells. A month or so before that date, the customer is notified by postcard to bring in his equipment. At that time, the equipment is given the standard series of performance tests for a modest fee. If the unit fails to meet one or more of the specifications, it could mean imminent trouble.

There's no question that high-quality test equipment is expensive, but it can pay for itself even if the level of business might not seem to justify the expense. My point is that there are ways of generating that business.

SIGNAL GENERATORS

There are quite a few types of signals employed in the servicing of modern hi-fi equipment. Most of those needed are available from a variety of sources for application in a controlled manner.

The AM band, not being hi-fi in nature, is usually a mere afterthought in the design of most FM stereo receiver. Because of this, and the inherent "low-fi" nature of the band, test equipment needs are minimal. In most cases, a so-called "service grade" rf signal generator will suffice quite nicely. It is also possible to get away with a simple crystal oscillator constructed of either transistors or IC digital logic gates and two crystals, one operating at 455 kHz and the other at 500 kHz. The lower of the two frequencies is the standard i-f in U.S.-made and most imported AM radio receivers. The higher frequency, 500 kHz, will produce usable harmonics at 1000 and 1500 kHz as well as the 500 kHz fundamental. These three frequencies are ideal for checking the low-, middle-, and high-end tracking on the AM broadcast band.

In some cases, the bench audio generator can also be pressed into service on the AM band. Many of these generators produce both sine- and square-wave outputs up to 100 kHz. If the generator is equipped to produce square waves, the output will be rich in harmonics considerably higher than the AM band.

The block diagram of a typical audio frequency signal generator is shown in Fig. 9-1. The basic method for generating the sine wave is the notch filter feedback system, also known as the "Wien bridge" circuit. In this application, an RC notch filter is inserted into the negative-feedback path of a differential amplifier. Since the filter attenuates those frequencies within its passband, they are not affected by the suppressive effects of the substantial amount of negative feedback used in the system. Consequently, the overall amplifier is a sharply resonant filter responding to the natural resonant frequency of the filter. If the gain is high enough, the circuit will oscillate to produce sine waves. Proper design will insure the purity of the sine wave curve.

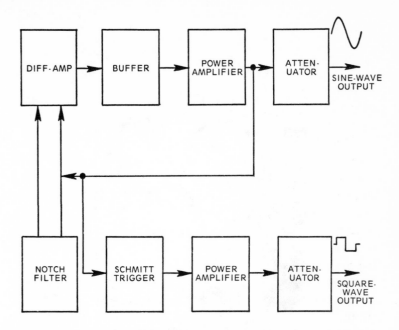

Fig. 9-1. Block diagram of a sine-and square-wave audio generator.

From the oscillator, the signal is fed to a buffer stage and then to a power amplifier. The buffer is necessary to prevent the oscillator frequency from varying due to changing load conditions. The usual practice is to lightly couple the oscillator into the buffer so that this effect occurs. The power amplifier builds up the signal level to a point where it can successfully deliver up to 10V ac (rms) to a standard 600-ohm load.

Figure 9-2 shows a typical sine-wave audio oscillator manufactured by Hewlett-Packard. In the block diagram of our hypothetical generator (Fig. 9-1), we see several additional stages that are needed to produce a square wave. A sample of the sine wave is taken from the power amplifier and is delivered to a circuit called a "Schmitt trigger." This is a two-state, or binary, circuit that changes state (on or off) only when a certain predetermined voltage level is present across the input terminals. When the instantaneous value of the sine wave reaches that point, the Schmitt trigger turns on. It remains on until the sine wave passes its peak value and once again falls back down to the trigger value (at least in theory). In some circuits there is a difference between t∕on and t∕off.

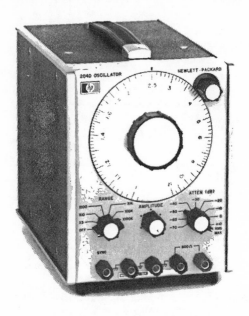

Fig. 9-2. Photograph of the Hewlett-Packard Model 204D audio generator.

This difference is called the circuit hysteresis. The less hysteresis the better. At that time, the Schmitt trigger turns off once again. The result from a properly designed circuit is a clean square wave generated from a sine-wave input.

The square-wave section of most generators has its own independent power amplifier and attenuator. Since rms meter readings don't mean much on square waves, this function is usually unmetered. The sine-wave attenuator is usually calibrated in either volts rms or dBm. In many cases, both systems are provided. Do not try to read the square-wave amplitude with any instrument other than an oscilloscope.

The FM portion of most stereo receivers contains somewhat more precise circuitry than the AM portion. Add to this the inherent precision required of any FM alignment technique, and the need for a quality FM generator is obvious.

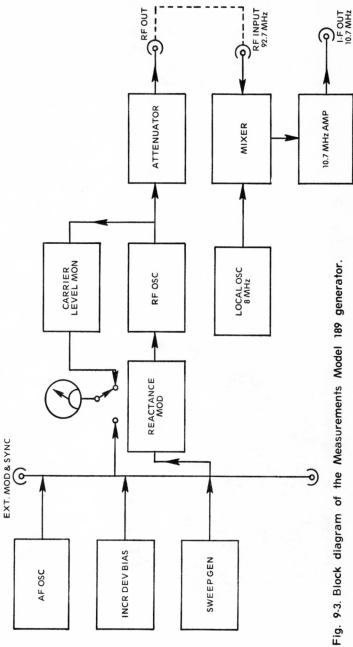

Fig. 9-3. Block diagram of the Measurements Model 189 generator.

If you elect to have separate FM and multiplex generators, there are several alternatives open in the selection of the right instrument.

One popular design, the Measurements Model 189, is shown in Figs. 9-3 and 9-4. It produces both FM i-f and front-end signals. The circuitry includes a reactance-modulated rf oscillator operating in the range covering the VHF FM broadcast band. The user can select either an audio tone or a swept signal to feed the reactance modulator. A 50-ohm rf attenuator at the output allows precise setting of signal level. The i-f function, sometimes needed for FM alignment, is derived in a heterodyne mixer circuit. An 82 MHz signal from a crystal oscillator beats against a 92.7 MHz signal from the main rf portion of the generator to produce the needed i-f of 92.7 minus 82 or 10.7 MHz. One disadvantage of this system is that precise FM i-f signal levels cannot be obtained merely by adjusting the attenuator dial. This isn't too much of a problem on i-f alignment procedures, however.

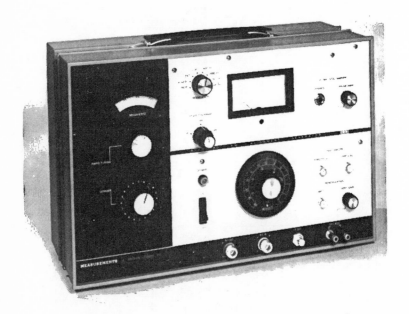

Fig. 9-4. Photograph of the Measurements Model 189 generator.

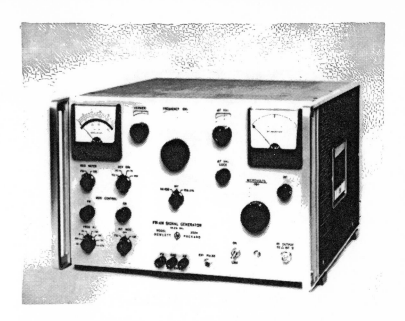

Fig. 9-5. Photograph of the Hewlett-Packard Model 207H generator.

Another FM signal source is shown in Figs. 9-5 and 9-6. In this system, Hewlett-Packard uses a high-quality FM signal generator to drive an i-f converter dubbed the "Univerter." This H-P Model 202H signal generator is a direct descendant of the famous Boonton Model 202 series. Although somewhat larger than the Measurements 189, the Univerter not only converts the input signal to a lower frequency, but it senses the input level (set by the attenuator on the 202H) and it matches it at the Univerter output. This allows the operator to set precise levels of 10.7 MHz signal using the master attenuator on the 202H as the control.

A good service bench multiplex stereo generator should include several features. It should, for example, provide the option (on a selector switch) of left-only or right-only stereo signals, as well as monaural (L + R) and unmodulated stereo (19 kHz pilot only). Another desirable feature is a separate audio output jack if the generator includes a provision for supplying the multiplex signal to an FM oscillator. That FM oscillator isn't strictly necessary, however, if the generator is to be used with an external lab-grade FM signal generator that

Fig. 9-6. Photograph of the Hewlett-Packard Model 207H Univerter.

can be externally modulated. In most cases, by the way, this will result in a more controlled and precise signal for any service procedure.

This author has had occasion to test a number of multiplex and FM generators over the past several years. A surprising number of the lower-cost types leak as much rf around the cabinet flanges as that which comes through the rf output jack at maximum attenuator settings. This rf leak is more than sufficient in most cases to drive the receiver being aligned into hard limiting. Not exactly the best circumstance for peaking up rf circuits!

BENCH METERS

There are two basic bench-meter types required for FM stereo receiver service. One is the familiar universal vtvm, while the other is a specialized version called the ac vtvm. The instrument pictured in Fig. 9-7 is a typical high-quality universal bench vtvm.

Since all new receivers are solid-state, it is necessary to use a vtvm with a set of low-value voltage ranges. Bias values and emitter-resistor voltage drops, both of interest in troubleshooting, are usually under 1V dc. Because of this, it is necessary to use a meter that allows nearly a full-scale reading of these low values. The top range need not exceed 300 volts.

Some technicians may prefer the newer digital readout multimeters. These instruments do have their uses in general troubleshooting; but for alignment and making dc adjustments, most technicians of this author's acquaintance still prefer the old-fashioned analog multimeter.

The Hewlett-Packard Model 400F ac vtvm (Fig. 9-8) should not be confused with the universal type that may also include an "AC volts" function. The frequency response and the range of full-scale values are much greater on a true ac vtvm. Many stereo receiver manufacturers require that any shop seeking to handle their brand of equipment as a factory authorized warranty service center must have an approved ac vtvm and, of course, certain other test equipment. In choosing an ac vtvm, look for both wide frequency response (100 kHz or

Fig. 9-7. Photograph of the Hewlett-Packard Model 427A general purpose.

so) and a wide range of full-scale sensitivity. Most of these instruments are calibrated in both dBm and volts/millivolts. Both scales are useful, so try and get a meter so designed.

OSCILLOSCOPES

The oscilloscope is vitally needed by the servicer who handles high-quality FM stereo receivers. He simply cannot do a full complement of service work without this instrument. The specifications for a suitable scope seem to be open to some discussion. On one hand, we hear the thought that a 500 kHz scope is sufficient, since this is basically audio work. Higher frequencies, such as those encountered in the FM i-f amplifier, can be monitored with a suitable demodulator probe. Others, however, seem to feel that one of the higher-priced models with a 3 dB response out to 10 or 15 MHz is needed so that the FM i-f signal can be viewed directly if so desired. In either case, and there are good arguments on both sides, good models are available from a variety of manufacturers.

The Hewlett-Packard type shown in Fig. 9-9 is available in both benchtop and rack-mounted variations. Most modern

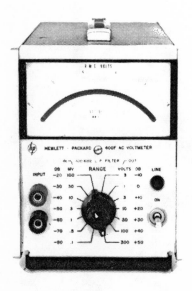

Fig. 9-8 Photograph of the Hewlett-Packard Model 400F ac. (Courtesy Hewlett-Packard.)

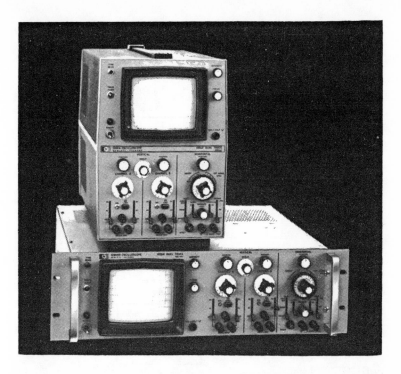
Fig. 9-9. Photograph of the Hewlett-Packard Model 1200A oscilloscopes.

technicians seem to prefer these triggered-sweep types over the older recurrent sweep models. When deciding whether to buy a scope with replaceable plug-in modules or a fixed vertical amplifier, you should consider the possible range of uses that will be made of the equipment. If none of the extra features offered by the plug-in vertical amplifier modules will ever be needed, it seems foolish to invest the substantial amount of money required to purchase one of those instruments.

One necessary feature for an oscilloscope in audio work is sensitivity. The minimum vertical deflection sensitivity should be in the neighborhood of 10 mV per cm. With this sensitivity, you can see all but the weakest of signals encountered in audio service. Some of the finer magnetic cartridges produce signal outputs so weak that even a sensitivity of 10 mV per cm will prove insufficient. In cases where it is absolutely necessary to know the voltage output from a phono

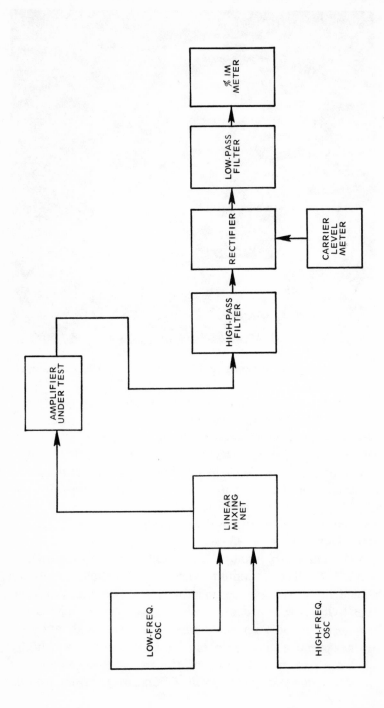

Fig. 9-10. Block diagram of the intermodulation distortion analyzer test setup.

cartridge, you can boost it with a calibrated gain instrument amplifier. These offer switch-selected gains of 20, 40, and 60 dB (A_V of 10, 100, and 1000 respectively). The need for such low-level measurements is so rare, however, that the extreme oscilloscope sensitivity might prove to be a poor investment.

Many newer oscilloscopes are available in dual-trace configurations. The audio technician may do well to consider one of these instruments. The dual-trace feature is useful when you want to monitor two channels in a stereo system or both the composite and pilot signal in the multiplex stages. Also, the phase relationship between the 38 kHz regenerated subcarrier and the pilot is easily checked with a dual-trace scope. This is an important adjustment in earlier sets, since it greatly affects the channel separation.

DISTORTION ANALYZERS

The Measurements Model 940 intermodulation distortion analyzer (Figs. 9-10 and 9-11) is one of the useful but, unfortunately, peripheral instruments on the audio bench. In an

Fig. 9-11. Photograph of the Measurements Model 940 IM analyzer.

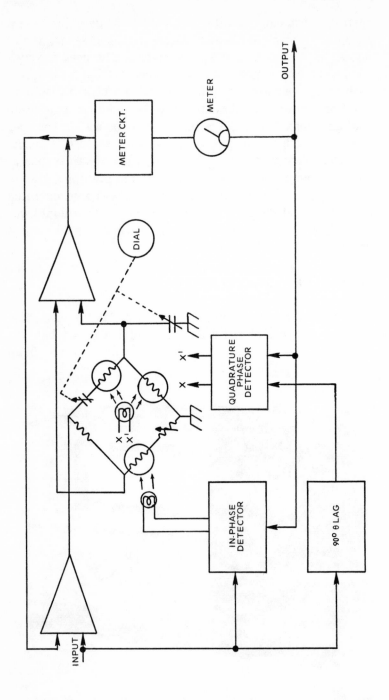

Fig. 9-12. Block diagram of the Hewlett-Packard harmonic distortion analyzer.

Fig. 9-13 Photograph of the Hewlett-Packard Model 334A harmonic distortion analyzer.

IM analyzer, two audio sine-wave signals, one high-frequency the other low, are mixed together in a linear network. Since the mixing takes place in a linear circuit, the low-frequency signal is merely superimposed on the high-frequency signal. Any nonlinearity, or distortion, is contributed by the amplifier under test. This causes the low-frequency signal to modulate (amplitude) the high-frequency signal. The IM analyzer determines to what degree, expressed as a percentage, this takes place in the amplifier or circuit under test.

The instrument shown in Figs. 9-12 and 9-13 is a harmonic distortion analyzer. The HDA is probably the most popular of the two types of distortion analyzers. A sine wave is the only waveform that is completely free of harmonic distortion. All other waveforms, including distorted sines, have some amount of harmonic content. The test for a true sine wave is to null out the fundamental frequency waveform, then test for the presence of remaining signals. If there are none, the waveform can be said to be truly sinusoidal.

In the HDA, the input signal from the amplifier or other circuit being tested is fed through either a null filter (most popular) or a high-pass filter. Any signals remaining at the output of the filter are harmonics, generated in the amplifier under test—assuming, of course, that the input sine wave is reasonably pure. If the instrument is calibrated so that the entire signal, before filtering, deflects the meter to exactly 100 percent (full scale), we get a direct reading of THD in percent when the filter is inserted. This gives a summation of all harmonic distortion products. A similar instrument, called a wave analyzer, is often used by audio engineers to seek the absolute amplitude of each individual harmonic. These latter instruments are of use to the service technician, but not important enough to justify their high price.

OTHER METERS

A number of other instruments are useful to the audio technician. Among these are the field strength meter, the digital frequency counter, and the wow and flutter meter.

The field strength meter is useful if your firm provides FM and TV antenna installation service in addition to stereo sales and service. It is useful but not essential.

The digital frequency counter is also handy. It can be used to calibrate shop test equipment. The author has seen one application where a DFC was left connected to the square wave output of an audio generator, while the sine-wave output was used in day-to-day servicing. The DFC offered a more accurate—although more expensive—frequency readout for the generator. The most important use for the DFC in audio work, however, is in setting the speed control on cassette tape players and recorders.

On the subject of speed measurement, it is interesting to note that an accurately recorded test tape offering a signal of exactly 1000 Hz will show a 10 Hz variation for each 1 percent of speed error. Therefore, a 1 Hz frequency variation will correspond to a 0.01 percent speed variation. Unfortunately, we must take any 1 Hz variation claims from a lower-priced instrument with a grain of salt. This is because all digital instruments are subject to "last digit bobble" (changing of the final digit). A frequency of 10,000 Hz would give better resolution, except that normal tape noise in that high end of the audio spectrum tends to obscure the true reading.

The wow and flutter meter is used to detect and measure the degree of frequency modulation added to recorded program material emanating from tape players (of all formats) and record players. These FM variations are caused by plus and minus periodic variations in the drive speed of the device under test.

MULTIPLE-FUNCTION AND SPECIAL-PURPOSE GENERATORS

A newer type of signal generator recommended by the national service departments of many stereo equipment manufacturers is pictured in Fig. 9-14. Several of these manufacturers have added this generator, the Sound Technology Model FM1000A, to their list of equipment required of warranty repair shops. This all solid-state instrument produces all of the signals needed to perform a normal alignment job on an FM stereo receiver. In addition, it also features a new type of alignment procedure called "dual sweep."

The Sound Technology "dual sweep" alignment response waveforms are shown in Fig. 9-15. In this system, a 10 kHz signal is superimposed on the normal 60 Hz signal used to frequency-modulate the VHF rf oscillator. With this arrangement, appropriate filtering techniques allow the user to view a very small portion of the discriminator S curve. This, in essence, displays the slope of the detector output curve on the screen of an oscilloscope. Interesting is the fact that the dual sweep actually displays a visual representation of the

Fig. 9-14. Self-contained FM rf and stereo composite signal generator, Sound Technology Model FM1000A. (Courtesy Sound Technology.)

intermodulation distortion produced by the receiver. The detector transformer secondary, or phase coil in quadrature detector circuits, is adjusted for the best symmetry of the curve shown in Fig. 9-15C.

Only two connections to the receiver are needed. The beauty of this system is that if the alignment points are accessible through holes in one of the cover plates, the receiver need not be opened for alignment. It is not necessary to find the appropriate scope connection point in the detector circuit

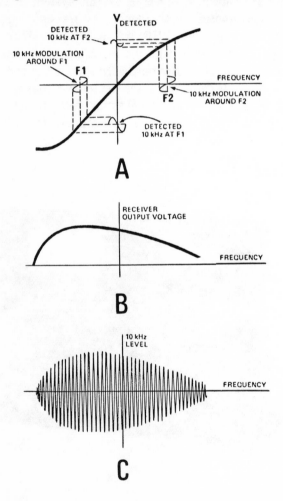

Fig. 9-15. Dual sweep waveforms used in the Model FM1000A alignment technique.

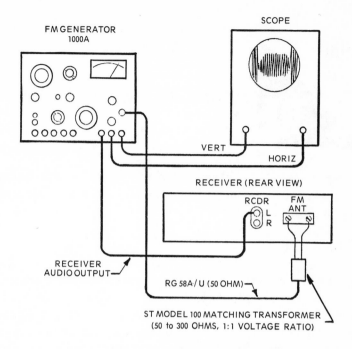

FM GENERATOR
1000A

SCOPE

VERT

HORIZ

RECEIVER (REAR VIEW)

RCDR
L
R

FM
ANT

RECEIVER
AUDIO OUTPUT

RG 58A-/ U (50 OHM)

ST MODEL 100 MATCHING TRANSFORMER
(50 to 300 OHMS, 1:1 VOLTAGE RATIO)

Fig. 9-1-6. Alignment setup using the FM1000A.

as is the case with the conventional alignment procedure. The
FM1000A system signal is applied to the antenna terminals of
the receiver and a sample of the detected audio output is
applied to a scope. In most receivers, best results are achieved
when this sample is taken from a tape recorder output or line-
driver jack (see Fig. 9-16). The utility and compactness of this
generator makes it extremely useful on a bench where high-
quality audio equipment is being serviced.

CHAPTER 10

Digital FM Stereo Tuner

On the market today, there is a highly advanced FM stereo tuner—the Heath Model AJ-1510. It employs a phase-locked loop design that merits attention.

FRONT END

The first requirement of a phase-locked loop FM front end (Fig. 10-1) is that the operating voltage must be controlled, particularly the local oscillator voltage. The output signal derived from the local oscillator is first scaled down to a lower frequency. This is done with a series of high-speed ac-coupled J-K flip-flops, usually of the ECL logic class. The output of this circuit it then fed to a programmable "divide-by-N" frequency scaler.

This type of circuit has a changeable frequency-division ratio. The counter can be preprogramed by external inputs to divide the input frequency by a specified ratio. The output of the divide-by-N counter is fed to a digital phase detector. The other input to the phase detector is derived from a crystal-controlled reference oscillator. In most cases, this oscillator operates at a frequency of 200 kHz, 100 kHz, or some sub-multiple of these frequencies.

The output of the phase detector is zero as long as the signals from the crystal oscillator and divide-by-N counter are coincident (in-phase). If a discrepancy exists, the detector will produce a dc voltage proportional to the difference. This dc voltage is fed through a filter that removes any residual rf on the "delta E" or "VCO tuning voltage" line going to the tuner front end. This voltage adjusts the VCO frequency until its divided-down counterpart is coincident with the signal from the reference oscillator. At that point, the detector output is zero volts and the influence on the VCO ceases. Thus, the VCO frequency remains constant. The secret to producing a specific frequency, corresponding to a specific FM channel, lies in the control of the divide-by-N counter. This circuit sees to it that only one frequency produced by the VCO will lock in on any particular division ratio. Tuning is accomplished by varying that division ratio.

The block diagram in Fig. 10-2 shows the specific PLL system used by Heath in the Model AJ-1510 FM stereo tuner. The basic PLL diagram is pretty much the same as the diagram in Fig. 10-1. In this case, however, the divide-by-N counter is fed by a circuit called a "preload decoder" and a data programer. The output of the preload decoder instructs the divide-by-N counter to operate in the modes listed on the chart in Fig. 10-3. It receives this information from the functional blocks shown in Fig. 10-4.

There are three means of applying instructions to the preload decoder. One of these is through the keyboard (see Fig. 10-5). In this mode, the operator punches in the frequency of the station he wishes to hear. If that station operates on 105.9 MHz, he simply pushes the buttons 1, 0, 5, and 9 in sequence. This causes the appropriate signal to program the divide-by-N counter.

Another source of data instructions is provided by a series of punched cards available from Heath. Each card is punched with the frequency of one of the user's favorite stations. The cards are inserted into a card reader that performs the function of programing the divide-by-N counter.

The third source of data for the divide-by-N counter is an "autosweep" circuit. In essence, this is a shift register circuit that "steps" down through the allowed division ratios, beginning at the high end of the FM band, until the presence of a signal stops the action.

The charts in Fig. 10-6 illustrate the number coding system used in the tuner data circuits. This is called the BCD-to-binary system. Decimal numbers (0 through 9) can be represented by binary digits if four spaces (bits) are provided. These are designated 1, 2, 4, and 8. Any decimal digit can be constructed with these four bits. The exact combinations used

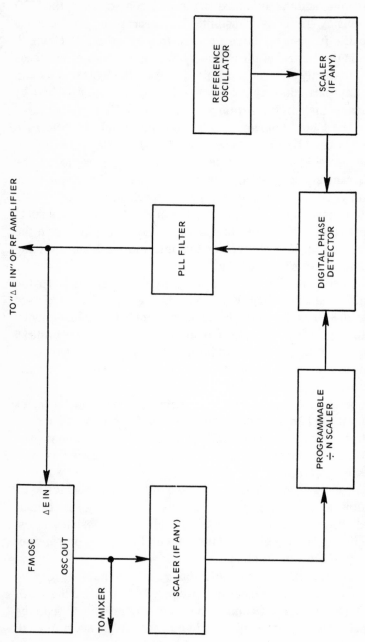

Fig. 10-1. Block diagram of a phase-locked loop local oscillator.

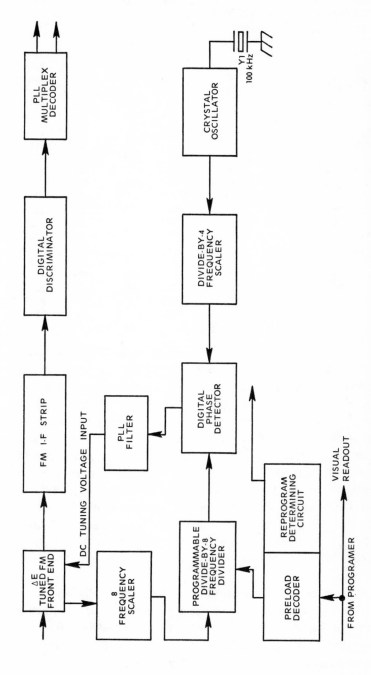

Fig. 10-2. Block diagram of the PLL local oscillator used in the Heath Model AJ-1510 tuner.

Station Frequency (MHz)	Preload Decoder Output	Station Frequency (MHz)	Preload Decoder Output	Station Frequency (MHz)	Preload Decoder Output	Station Frequency (MHz)	Preload Decoder Output
						100.7	213
88.1	150	92.3	171	96.5	192	100.9	214
88.3	151	92.5	172	96.7	193	101.1	215
88.5	152	92.7	173	96.9	194	101.3	216
88.7	153	92.9	174	97.1	195	101.5	217
88.9	154	93.1	175	97.3	196	101.7	218
89.1	155	93.3	176	97.5	197	101.9	219
89.3	156	93.5	177	97.7	198	102.1	220
89.5	157	93.7	178	97.9	199	102.3	221
80.7	150	93.9	179	98.1	200	102.5	222
89.9	159	94.1	180	98.3	201	102.7	223
90.1	160	94.3	181	98.5	202	102.9	224
90.3	161	94.5	182	98.7	203	103.1	225
90.5	162	94.7	183	98.9	204	103.3	226
90.7	163	94.9	184	99.1	205	103.5	227
90.9	164	95.1	185	99.3	206	103.7	228
91.1	165	95.3	186	99.5	207	103.9	229
91.3	166	95.5	187	99.7	208	104.1	230
91.5	167	95.7	188	99.9	209	104.3	231
91.7	168	95.9	189	100.1	210	104.5	232
91.9	169	96.1	190	100.3	211	104.7	233
92.1	170	96.3	191	100.5	212	104.9	234

Fig. 10-3. Chart of the preload decoder output, Heath Model AJ-1510.

are shown in Fig. 10-6A. The charts in Figs. 10-6B and 10-6C show how four four-bit numbers can be used to generate the various FM frequencies loaded into the divide-by-N counter. The four columns are designated here as A, B, C, and D. Since the column A data is either a decimal 1 or decimal 0, this is connected to a switch. It becomes 1 for all FM band frequencies above 100 MHz. The remaining columns (B, C, and D) must be selected in a slightly more complex manner. In the example shown, the selected frequency is 107.9 MHz. This is selected by making column A equal to 0001, column B equal to 0000, column C equal to 0111 (decimal 4 + 2 + 1 equals 7) and column D equal to 1001 (decimal 8 + 1 equals 9). This is read as 0001 0000 0111 1001, or in decimal 107 9. The decimal point is understood.

An example of the selector card is shown in Fig. 10-7. The two possible binary states of 1 and 0 are indicated by either a cutout or blinded part of the card edge. Positions on the card edge are: A1, B8, B4, B2, B1, C8, C4, C2, C1, D8, D4, D2, and D1. In the example shown in Fig. 10-7, a frequency of 98.7 MHz is

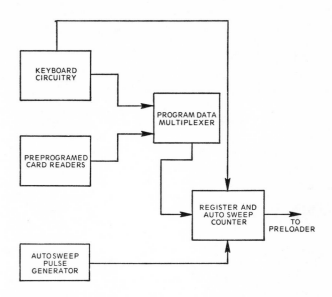

Fig. 10-4. Block diagram showing the data sources used to program the AJ-1510 local oscillator PLL.

selected by cutting out B8, B1, C8, D4, D2, and D1. In binary, this represents 0000 1001 1000 0111, and 098 7 in decimal. An optical card reader picks up this information and feeds it to the programing circuits.

A schematic of the Heath AJ-1510 FM tuner and the divide-by-N scaler is shown in Fig. 10-8. Notice that all of the tuned circuits in this diagram are voltage-tuned. A sample of signal from the FM local oscillator is fed to the scaler through a common-base buffer amplifier. Also used as buffers are several ECL ac-coupled logic gates. Each of the ECL flip-flops following the buffers is capable of reducing the input frequency by a factor of two. The output frequency in each case is one half the input frequency. The actual total division ratio is $\frac{1}{2}$ x $\frac{1}{2}$ x $\frac{1}{2}$ equals $\frac{1}{8}$. An additional pair of gates couples the signal to the remainder of the tuner.

From the scaler, the signal is passed to the divide-by-N counter (see Fig. 10-9). Here it is further scaled down by a division factor determined by the programing circuitry. The output of the divide-by-N counter is fed to a phase detector circuit (see Fig. 10-10). The other phase detector input is fed

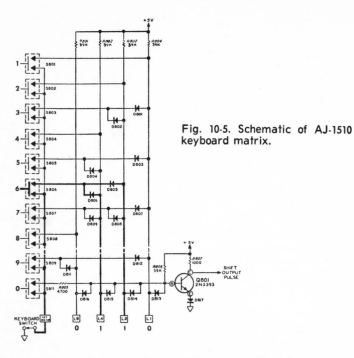

Fig. 10-5. Schematic of AJ-1510 keyboard matrix.

by the 25 kHz reference signal. This signal is developed from a 100 kHz crystal oscillator fed through a pair of cascade J-K flip-flops. (Divide-by-2 + divide-by-2 equals divide-by-4. 100 kHz—4 equals 25 kHz.) The output of the phase detector is a dc voltage whose absolute level is determined by the coincidence, or lack of it, between the reference pulse and the pulse from the output of the divide-by-N counter.

The stations frequency readout uses digital display tubes. The readout and readout driver circuits are shown in Fig. 10-11. In this arrangement, integrated circuits, designed to operat as "BCD-to-seven-segment-decoders," receive the A, B, C, and D inputs from the programing circuits and convert this to the appropriate seven-segment display. Both the IC drivers and the display tubes are of the RCA type.

I-F AND DETECTOR

The FM i-f amplifier strip and FM detector in this tuner are just as advanced as the rest of the unit (Fig. 10-12). The

178

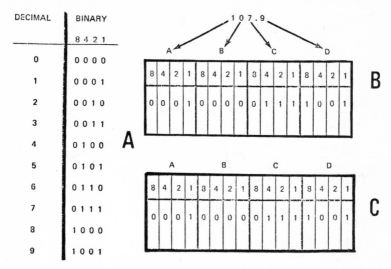

Fig. 10-6. Charts showin the BCD-to-decimal coding used in the AJ-1510 tuner.

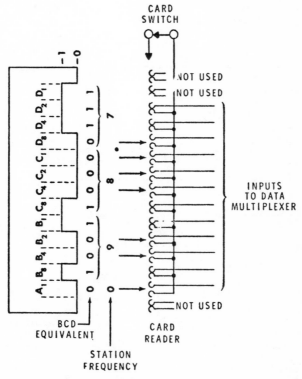

Fig. 10-7. Schematic of the BCD card reader. (Courtesy Heath Co.)

179

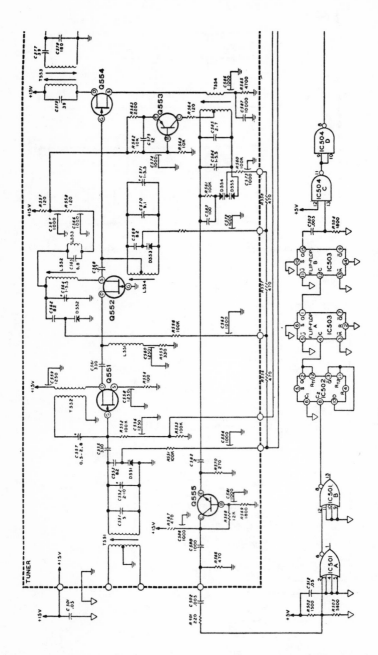

Fig. 10-8. Schematic of the voltage-tuned FM tuner subassembly.
(Courtesy Heath Co.)

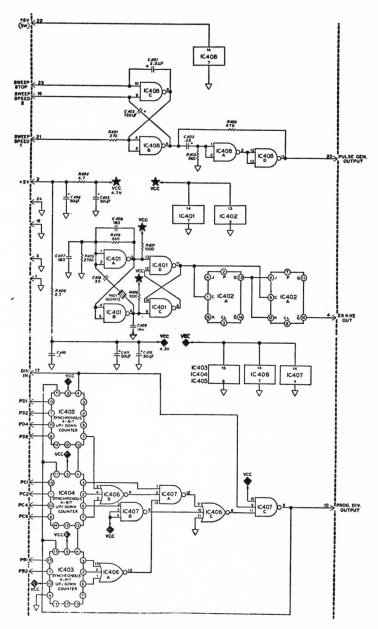

Fig. 10-9. Schematic of the reference oscillator and divide-by-N counter.
(Courtesy Heath Co.)

181

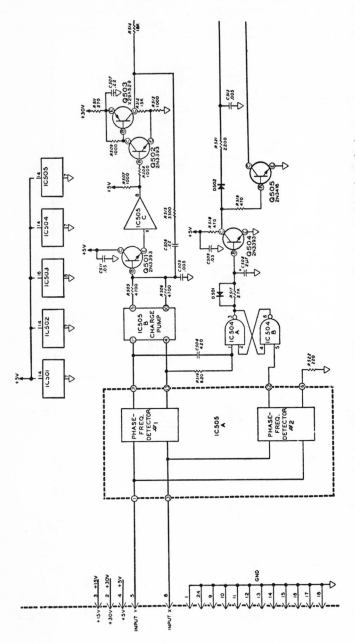

Fig. 10-10. Schematic of the frequency-phase detector used to phase lock the local oscillator to the reference oscillator. (Courtesy Heath Co.)

FM i-f output from the voltage-controlled PLL tuner is first fed to an amplifier stage that uses the well known uA703A IC (IC602). This IC is a differential amplifier especially designed for use as an FM i-f amplifier. Further amplification is provided by another pair of integrated circuits. These are Motorola's type MC1357P. Selectivity in this section is provided by a pair of computer-designed LC bandpass filters designated as F601 and F602.

The FM detector is a digital type. A block diagram is shown in Fig. 10-13. The 10.7 MHz signal delivered by the FM i-f amplifier is extremely well limited. This means that the waveshape of the individual cycles is roughly square. These, in essence, are pulses.

The first stage in the detector is an IC retriggerable monostable multivibrator. This IC is designated IC604 in Fig. 10-12. The monostable triggers on every time an input pulse is delivered by the FM i-f amplifier. The nature of a monostable is such that it has but one stable state. When triggered by an external pulse, it switches to the unstable state for a period of time determined by the circuit component values. At the end of that time, it reverts back to its stable state,where it remains an indefinite period of time until retriggered. The pulse length produced at the output of the monostable is constant and is less than the period of the 10.7 MHz FM i-f input signal. The output of the monostable, then, is a series of pulses of constant amplitude and width. What varies with the FM modulation is the number of pulses produced. As the signal is varied by the audio modulation, either more or fewer pulses are produced by the monostable.

The output pulses produced by the monostable are fed to a pair of integrators. Those pulses, by the way, are produced in two equal but opposite-phase pulse trains designated as Q and $\bar{Q}$ (read "not Q"). The two integrators in the circuit in Fig. 10-12 consist of resistors R619, R628, R627, R629 and capacitors C610, C637, C638, and C640. The integrator produces a dc voltage level that is proportional to the spaces between the monostable output pulses. Since the space width varies with the number of pulses produced per unit of time, this is proportional to the audio signal modulating the FM signal.

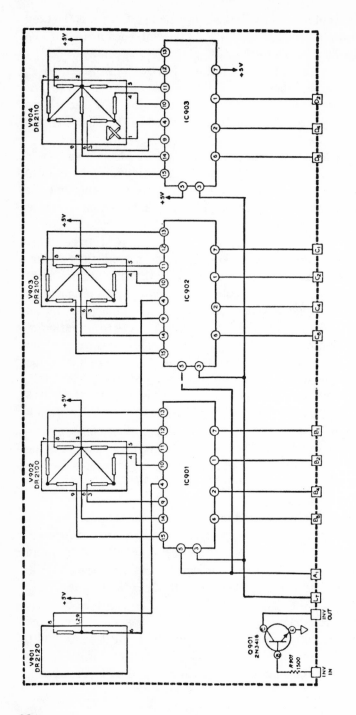

Fig. 10-11. Schematic of the circuits used to provide a digital readout of the tuned frequency. (Courtesy Heath Co.)

Because the outputs from the monostable are out of phase, the outputs from the two integrators are also out of phase. These two signals are fed to the two opposite polarity inputs of operational-amplifier IC605. Here, they are first summed and are then amplified. Since the two op-amp inputs are of opposite polarity, the circuit tends to cancel noise signals. That noise tends to appear at both inputs equally, a fact that makes them cancel each other. This circuit, therefore, allows a high signal-to-noise ratio.

MULTIPLEX DECODER

Another phase-locked loop is used in the composite multiplex decoder circuit. This stage takes the composite signal delivered by the FM detector and decodes it into discrete left- and right-channel audio signals. The circuit of the PLL stereo decoder is shown in Fig. 10-14. This circuit uses the Motorola "coilless" stereo decoder IC. The circuit offers superior performance in some respects because the internal oscillator will phase lock only on the 19 kHz pilot signal transmitted by the broadcast station. This gives the PLL decoder a high degree of rejection to SCA and other unwanted signals without the need for expensive traps. The only adjustment is a 5K potentiometer, which is turned until phase lock is produced. This is only mildly critical because the PLL's nature is to reach out and grab for lock much in the manner of afc.

MUTE

The circuit in Fig. 10-15 shows the diode switching circuit used in the squelch circuit (mute) of the Heath AJ-1510. Diode switching takes advantage of the fact that solid-state diodes will pass an ac signal without rectification, provided that it is superimposed on a dc level that is, relatively speaking, much larger than its own value.

Transistor Q2 is normally saturated, which provides a ground return for the diode cathode circuits. The base of

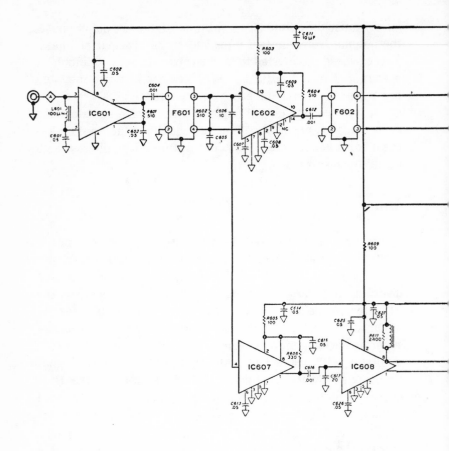

Fig. 10-12. Schematic of the i-f amplifier and digital detector.

transistor Q1 is normally in the low or ground state when the receiver output is unsquelched. The squelch signal drives this point high, which forward biases Q1. When Q1 begins to drive toward saturation, Q2 begins to cut off. This causes the collector voltage at Q2 to rise, approaching the V_{cc} level of 15V. This voltage is also coupled to the common cathodes of diode pairs DS1 and DS2. With a high voltage on the common cathode, these diodes will not pass signal and will be, in fact,

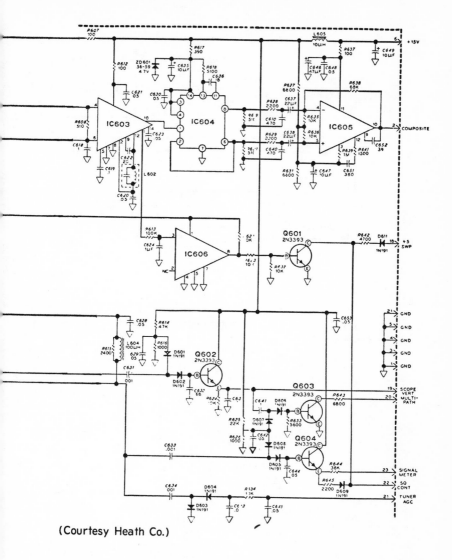

(Courtesy Heath Co.)

cut off. Diode pair DS3 shunts the two audio lines. When the receiver is unsquelched, the collector of Q1 is high and this diode pair is cut off. When the squelch signal is present, however, the collector of Q1 goes low (less than 0.3V dc), allowing a ground-return path for the common cathodes of DS3. This, in effect, places a low impedance across the audio lines to ground and enhances the squelch function of the receiver.

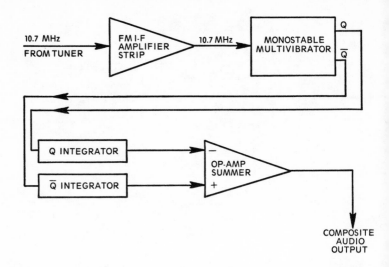

Fig. 10-13. Block diagram of the digital FM detector in the Heath Model AJ-1510.

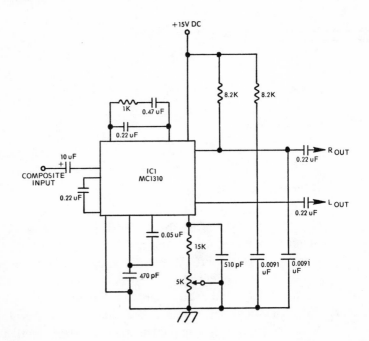

Fig. 10-14. Schematic of the Model AJ-1510 stereo decoder using the Motorola MC-1310P IC.

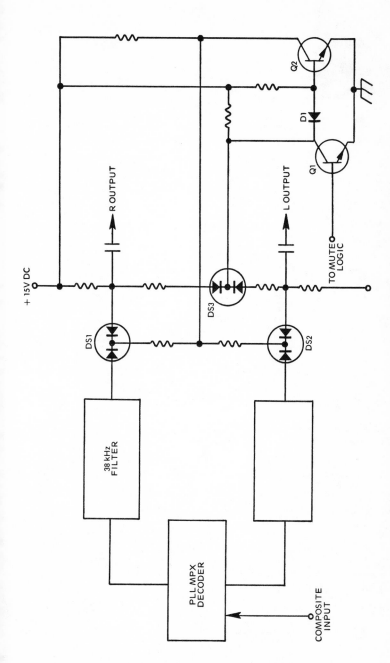

Fig. 10-15. Schematic of the diode switching mute circuit used in the AJ-1510.

189